SUPER GOLFONOMICS

SUPER GOLFONOMICS

Stephen Shmanske
California State University, USA

NEW JERSEY • LONDON • SINGAPORE • BEIJING • SHANGHAI • HONG KONG • TAIPEI • CHENNAI

Published by

World Scientific Publishing Co. Pte. Ltd.
5 Toh Tuck Link, Singapore 596224
USA office: 27 Warren Street, Suite 401-402, Hackensack, NJ 07601
UK office: 57 Shelton Street, Covent Garden, London WC2H 9HE

Library of Congress Cataloging-in-Publication Data
Shmanske, Stephen, 1954–
Super golfonomics / Stephen Shmanske.
pages cm
Includes bibliographical references and index.
ISBN 978-9814612548 (alk. paper)
1. Golf courses--Economic aspects. 2. Golf--Economic aspects. 3. Professional sports--Economic aspects. 4. Economics--Psychological aspects. I. Title.
GV975.5.S56 2015
796.352068--dc23

2014018385

British Library Cataloguing-in-Publication Data
A catalogue record for this book is available from the British Library.

In-house Editor: Lum Pui Yee

Printed in Singapore

This book is dedicated to:

Tina Stevens

Shadow, Gypsy, Sable, Scout, and Cinder

Preface

In *Golfonomics* I enthused about how happy I was pursuing two of my passions, golf and economics, at once, and even getting paid to do so. Now, a decade later, I affirm that I have not changed in this regard. I feel lucky, even blessed, that I was able to combine a love of sport in general and golf in particular with a career as an economist. I discovered sports economics at just the right times in both my career and in the burgeoning field of sports economics, getting in on the ground floor and pioneering the economic analysis of golf's data and institutions. The rest, as they say, is history. Additionally, it has just been plain fun.

There are many people to thank for helping me in the process of writing this book–parents, family, (even pets), friends, teachers, mentors, classmates, colleagues, students, golf partners, coaches, and teammates. I extend gratitude to all of these. I am lucky that gratitude is not subject to the usual budget constraint, if it were I would become a poor man indeed.

Many of the chapters that follow report on previously published research and specific thanks can be given to journals, editors, publishers, readers, and conference participants associated with each of the papers. These will be named below. There is also a sense in which all my previous published and unpublished research laid part of the groundwork for the specific research included in this book. For these unnamed readers and conference participants I am also thankful.

Chapter 2 is mostly a reprint of Chapter 8 from *Golfonomics*.[1] I thank the editors and publishers at World Scientific Publishing Company. This

[1]See, Shmanske, S. (2004a). *Golfonomics*. (World Scientific Publishing Co., Inc. River Edge, NJ).

material was never published in a journal but was presented in a workshop at California State University, East Bay (CSUEB, formerly C. S. U., Hayward) and at the Western Economic Association, International (WEAI) conference in San Francisco, July 2001. I thank the organizers of the WEAI session, Larry Hadley and Elizabeth Gustafson, and discussant, Stacey Brook.

Chapter 3 is comprised of textbook level intermediate theory and has not been previously published. Helpful comments were received from Tina Stevens on the original draft.

Chapter 4 draws directly from two published articles.[2] I thank the referees and editors of *Public Choice* who published the theoretical model showing how pricing below cost by a governmental agency can actually lead to fewer firms with lower quality and lower quantity. Charles Baird, Leo Kahane, Dwight Lee, and M. Bruce Johnson also made valuable comments on early drafts. I acknowledge the financial support of the Smith Center for Private Enterprise Studies and the helpful suggestions received at the CSUEB economics workshop. The empirical test of the proposition was published in the *International Journal of the Economics of Business* and I thank the referees and editors especially, H. E. Frech III, and Tina Stevens for helpful comments. The empirical results were also presented in a workshop at CSUEB and at the WEAI conference in Seattle, July 2002. My gratitude goes to the organizers of the WEAI session, Larry Hadley and Elizabeth Gustafson, and discussant, Stephen J. Spurr.

Chapter 5 draws from material published in *The Atlantic Economic Journal*.[3] I thank the editors and the anonymous referees for their input and suggestions. I also received assistance from Leo Kahane. A version of the research was also presented in the workshop at CSUEB. Since the workshop at CSUEB has been acknowledged several times already, now may be a good time to thank its audience which consists of a different

[2]See, Shmanske, S. (1996). Contestability, Queues, and Governmental Entry Deterrence, *Public Choice*, 86, pp. 1-15, and Shmanske, S. (2004b). Market Preemption and Entry Deterrence: Evidence from the Golf Course Industry, *International Journal of the Economics of Business*, 11, pp. 55-68. The latter is a Taylor & Francis Group journal, see www.tandfonline.com

[3]See, Shmanske, S. (2007). Consistency or Heroics: Skewness, Performance and Earnings on the PGA TOUR, *The Atlantic Economic Journal*, 35(4), pp. 463-471.

group of students each academic year along with a smattering of my colleagues at each seminar depending upon their own class schedules. More or less regular attendees that I'll acknowledge by name include, Charles Baird, James Ahiakpor, Greg Christainsen, Leo Kahane, Nan Maxwell, Tony Lima, Adrian Stoian, and Jed DeVaro.

Chapter 6 extends the themes from Chapter 5 in a paper published in the *Journal of Sports Economics*.[4] I thank the editor, Leo Kahane, and the unnamed referees for their suggestions. I also appreciate the feedback received at the WEAI conference in Seattle, July 2007, in a session organized by David J. Berri, Brad R. Humphreys, and Anthony C. Krautmann with discussant, Thomas A. Rhoads. This research was also presented at the CSUEB workshop.

Chapter 7 follows a line of research that was originally published using 1998 data in the *Journal of Sports Economics* and fleshed out in *Golfonomics*.[5] The material was then further extended with data from 2008 and published in Chapter 3 of *The Oxford Handbook of Sports Economics, Volume 2*.[6] Thanks go to my co-editor, Leo Kahane, and to the anonymous reviewers all around. I appreciate the comments offered by Tina Stevens on the final product. The earlier material was presented in the CSUEB workshop and at the WEAI meetings in Vancouver, July 2000 in a session organized by Larry Hadley and Elizabeth Gustafson with special thanks due to Michael Ransom and Nan Maxwell.

[4]See, Shmanske, S. (2008). Skills, Performance, and Earnings in the Tournament Compensation Model: Evidence from PGA TOUR Microdata, *Journal of Sports Economics*, 9(6), pp. 644-62.

[5]See, Shmanske, S. (2000). Gender, Skill, and Earnings in Professional Golf, *Journal of Sports Economics*, 1(4), pp. 385-400, and Shmanske, S. (2004a). *Golfonomics*. (World Scientific Publishing Co., Inc. River Edge, NJ).

[6]See, Shmanske, S. (2012b). *The Oxford Handbook of Sports Economics, Volume 2: Economics through Sports*, eds. Shmanske, S. and Kahane, L. H., Chapter 3 "Gender and Discrimination in Professional Golf," (Oxford University Press, Inc., New York) pp. 39-54. By permission of Oxford University Press, USA.

Chapter 8 was published in part in Chapter 4 of the *Handbook on the Economics of Women's Sports.*[7] My gratitude goes to co-editors, Eva Marikova Leeds and Michael A. Leeds, who encouraged me to undertake the research and offered many helpful, guiding comments. I also appreciate the feedback from the audience at the CSUEB workshop.

Chapters 9 and 10 report and extend research originally published in the *International Journal of Sports Finance.*[8] I thank the editors and referees who offered helpful suggestions. Special thanks go to Leo Divinagracia. Without his computer programming assistance the paper would not have been completed. The paper was also presented in the CSUEB workshop. An earlier version of the paper was presented at the International Association of Sports Economists meeting in Gijon, Spain, May 2000. Extra thanks go to Placido Rodriguez, Stefan Kesenne and Jaume Garcia who organized the conference and hosted my visit, and to discussant, Kevin Quinn. The paper was also presented at the WEAI meetings in Honolulu, July 2008 in a session organized by David J. Berri, Brad R. Humphreys, and Anthony C. Krautmann, with discussant Thomas A. Rhoads.

Chapter 11 is based on research published in the *Journal of Economics and Finance.*[9] I especially thank the editors of the special symposium on sports gambling, Rodney Paul and Andrew Weinbach. An early version of the paper was presented in the workshop at CSUEB. It was also presented at the WEAI meetings in Vancouver, July 2004, in a session organized by David Berri, Elizabeth Gustafson, and Larry Hadley with discussant, Michael A. Leeds.

[7]See, Shmanske, S. (2013). *Handbook on the Economics of Women's Sports*, eds. Leeds, E. M. and Leeds, M. A., Chapter 4 "Gender and Skill Convergence in Professional Golf," (Edward Elgar Publishing Ltd, Northampton, Massachusetts) pp. 73-91.

[8]See, Shmanske, S. (2009). Golf Match: The Choice by PGA Tour Golfers of Which Tournaments to Enter, *International Journal of Sports Finance*, 4(2), pp. 114-135.

[9]See, Shmanske, S. (2005). Odds-Setting Efficiency in Gambling Markets: Evidence from the PGA TOUR, *Journal of Economics and Finance*, 29(3), pp. 391-402.

Chapter 12 was published in Chapter 25 of the *International Handbook on the Economics of Mega Sporting Events*.[10] My gratitude goes to the co-editors, Wolfgang Maennig and Andrew Zimbalist, who made helpful comments and pushed me to consider additional tests. I am also indebted to Jed DeVaro for technical assistance and advice. Finally, in what has become a habit, I thank the participants in the CSUEB economics workshop.

On the less academic side I am grateful to all my golf buddies and partners over the years. They have taught me the game and/or listened to my academic soliloquies. Many were simply casual pickups at a variety of golf courses around the world. Others were occasional or regular partners. In more or less chronological order thanks to my father, Bernard F. Shmanske, my brothers, Bernard F. Jr., Matthew, and Clement, my college friends, Dave Lidstone and Scott Specht, my mother, Elizabeth A. Shmanske who picked up the game when the nest finally emptied, economist golfers, Greg Christainsen, and Joe Fuhrig, and all around sports and golf enthusiast, Ed Martin.

Close in-laws and friends have sustained me and my family more than they know and in more ways than can be said. Thanks Jeanne Stevens, Jim Stevens, Rosann Greenspan, and Lisa Zemelman. As far as my immediate family, I could not have done it without the calming and distracting influences of Sable, Scout, and Cinder. Finally, there has been the constant love and guidance of Tina Stevens, for better or worse (definitely better), for richer or poorer (richer in so many ways), until death do part us.

Stephen Shmanske

[10]See, Shmanske, S. (2012a). *Handbook on the Economics of Mega Sporting Events*, eds. Maennig, W. and Zimbalist, A., Chapter 25 "The Economic Impact of the Golf Majors," (Edward Elgar Publishing Ltd, Northampton, Massachusetts) pp. 449-460.

Contents

Chapter 1

Introduction

I remember when I first started research on the economics of golf. It was 1987 and I was sick with the flu. My wife, Tina, bought me a golf magazine to read and it had a spreadsheet of statistics for the top 60 money winners on the PGA TOUR for the 1986 season. Primitive now, but fantastic for the time, software allowed me to run regressions on my ten-megabyte hard drive desktop behemoth to derive the earliest production function for professional golf. It didn't even seem like work. I cannot remember as vividly the genesis of each particular line of research that I have pursued since that first paper. But it seems to me now that, like a putt from the top shelf of a two-tiered green, my research started rolling slowly and picked up speed as I moved from one golf topic to another, enjoying it every bit of the way.

By the early 2000's, I had collected enough material to bring forth *Golfonomics* which was published in 2004. But my research was just starting to heat up. Since the material in *Golfonomics* was finalized, I have published ten more pieces in a variety of places. Additionally, other researchers have started to expand upon my work and open up new lines of inquiry. It is time for some more *Golfonomics*. In fact, *More Golfonomics* was the working title of this book until I decided to do some coattail riding and entitle this volume, *Super Golfonomics*.

The avid reader and follower of popular economic culture may remember the best selling book by Steven Levitt and Stephen J. Dubner, *Freakonomics*, that was published in 2005, the year after *Golfonomics*. *Freakonomics* applied economics and econometrics to problems that seem unconnected to economics, showing how economic explanations could shed light on phenomenon as varied as abortion and crime, working conditions for crack cocaine dealers, and cheating by Sumo wrestlers. Although there is much to criticize in the book, it became a popular best seller. Its success can be attributed to the cleverness of its subject matter, its marketing and promotion by the publisher, and the combination of Levitt's insights as an Economics professor at the University of Chicago and Dubner's writing skills as a journalist with the *New York Times*. The book was so successful that they published a sequel, *Super Freakonomics*, in 2009.

At first I was happy telling everyone that *Golfonomics* was like *Freakonomics* except that all the examples were about golf, and . . . , that *Golfonomics* was first. I would also tell them that despite what sounds like a narrow focus, the examples in *Golfonomics* were more interesting and more compelling than those in *Freakonomics*. To list just three comparisons, which do you think is more interesting to members of the general public? (1) Tiger Woods' earnings and celebrity or the link between abortion and crime, (2) the potential for women to compete against men or Sumo wrestling, and (3) high corporate executive salaries as incentives or working conditions for crack cocaine dealers. In each of these comparisons the former topic is in *Golfonomics* and the latter is in *Freakonomics*. Unfortunately, my personal sales efforts did not have the intended result of making *Golfonomics* a million seller. I wracked my brain to find out why this was so. I originally suspected that my publisher's high-price/low-advertising-budget style was not as good as the low price strategy coupled with a massive promotional campaign chosen by the publishers of *Freakonomics*. Ya think? While I still suspect that this is the case, it is also possible that the professional writing capabilities of a major newspaper journalist might have something to do with it. Whatever the reason, Levitt and Dubner had the best seller and I didn't. It might have been nice to claim that Levitt and Dubner stole the idea for their title from me, but I doubt that it was the case. Anyway, you can't judge a book by its cover. I end up jokingly proclaiming that since they rode on my coattails with *Freakonomics*, it is now my turn to free ride, hence the title of this volume, *Super Golfonomics*.

Incidently, several sports economics colleagues of mine who will remain nameless here have suggested that the title, *Golfonomics*, did get them to be interested in the book in the first place. I find this gratifying. I hope they liked it enough to also be interested in *Super Golfonomics*. I also like the fact that since it is a made up word, computer search engines bring my work, or references to it, up to the top of the page very quickly. Finally, they say that imitation is sincere flattery. And I am very flattered that one of the preeminent sports economists and my old world style namesake, Stefan Szymanski, has used a spinoff of the idea in his own book. While we are not too far removed from the topic of coattail riding, I am happy to say that more than once I have been asked if I was spelling my name differently, the questioner thinking that some of Stefan's work was actually mine. There is definitely a positive externality that rubs off on me; I hope that not too many negative spillover effects go the other way. To me, Stefan is a workaholic, possibly attributable to the fact that he is a self-proclaimed "worst athlete in the world." Stefan, therefore, does not waste his time on the golf course like I do and is a most prolific scholar. Football, or in his concession to the American market, soccer, is his favorite and I wish him great success with his book, *Soccernomics*.[1]

Enough about the title, what's actually on the inside? There really is something for everyone, both in terms of the topics covered and in terms of areas of economics employed in the analysis. Topics include gender discrimination, policy, and performance; posted odds gambling; golf's Major tournaments; competition, entry, and contestable markets; advertising; how to win tournaments; which tournaments to enter; and slow play on the golf course. All of these topics have some relationship to golf or golf statistics, but many have aspects that are relevant to society as a whole. With this in mind, two organizing themes that appeared in *Golfonomics*, and in other work I have done in sports economics[2] are also

[1]The book is coauthored with a longer full title, see, Kuper, S. and Szymanski, S. (2009) *Soccernomics: Why England Loses, Why Spain, Germany, and Brazil Win, and Why the US, Japan, Australia, Turkey-and Even Iraq-Are Destined to Become the Kings of the World's Most Popular Sport*, (Nation Books, New York).

[2]See Kahane, L. H. and Shmanske, S. eds. (2012) *The Oxford Handbook of Sports Economics, Volume 1: The Economics of Sports*, (Oxford University Press, Inc., New York), and Shmanske, S. and Kahane, L. H. eds. (2012) *The Oxford Handbook of Sports Economics, Volume 2: Economics through Sports*, (Oxford University Press, Inc., New York).

evident in this book. First, some of the book is better described as the use of economics to examine and understand various aspects of the golf industry, for example, the use of inventory modeling to understand slow play on golf courses. Second, some of the book is better described as the ability to exploit statistics developed in sports settings to address larger social problems, such as gender discrimination or the unintended consequences of market intervention. Within golf itself, some chapters will interest fans, others will interest professional golfers and coaches, others will interest golf course builders and investors, and most will interest the typical enthusiastic amateur golfer. Any golfer who likes to read, and who wants a golf book with some intellectual heft, will enjoy *Super Golfonomics*.

With respect to economics and economists there is also much variety. Areas of economics and related fields that are called upon include labor economics, finance and efficient markets, public finance, the economics of discrimination, econometrics, statistics, production theory, contestability and entry, inventory management, and growth. The basics of whatever theories are used are presented and explained in everyday language, however, so the non-economist should not fret. Indeed, I hope that one result of reading the book might be a slightly increased understanding of economics and economists by those unfamiliar with the economic way of thinking.

In the chapters that follow, sometimes past results are confirmed and strengthened with new statistical analysis, as in the cases of gender-based earnings decompositions, the efficient setting of betting odds, and the lackluster effects of staging a mega event. In other cases new applications of existing theory are pursued, as in the application of inventory control to golf course pace of play, the statistical ability to examine the variance (and skewness) along with the mean of skill distributions, or in ferreting out the chicken/egg problem of prodigiously long drives stimulating fan interest leading to high purses versus high purses supplying incentives for intense practice leading to longer and longer drives. I hope that economics students, economists in a variety of fields, and especially sports economists will find the book interesting, in some cases provocatively so, and will engender new ideas for future research.

Although each chapter is a self-contained whole, there are connections among groups of chapters. Chapters 2-4 deal with slow play and other effects of waiting on the golf course or waiting for a tee time. As a subset of the economics of golf, these chapters could be called golf course economics. Chapters 5-10 deal with a variety of aspects of the economics of professional golf. Among them are the relationships between skills, scoring, and earnings, an examination of gender differences between the PGA TOUR and the LPGA, and the decisions of professional golfers about which tournaments to enter. Through Chapter 10, the economic topics that have been touched upon include, demand and supply, rationing by waiting, inventory modeling, production functions, earnings functions, the economics of discrimination, and labor economics. The remaining three chapters add to this mix the economics of gambling, efficient markets theory, local public finance, the economics of advertising, and behavioral economics. Some readers may choose to read from cover to cover while others will want to jump to topics of most direct interest to them. The following chapter by chapter overview introduces the topics for readers of either type.

Chapter 2, "The Economics of Slow Play," is the only complete overlap with material from the original *Golfonomics*. The chapter has a few additions but is largely the same as before. My choice here is to give the community of golfers and golf course operators an additional chance to understand and fix one of the most important causes of slow play, or more accurately, long rounds of golf. Many of us have had the experience of playing behind (or several groups behind) a group of golfers that is not keeping up with an appropriate pace of play. This can happen at any course at any time. However, when the same pattern of slow play occurs time after time on a particular course (yours) at a particular time (essentially, the later the worse) or on particularly crowded days (like weekends), the cause has more to do with golf course design and golf course operations than it does with individual slow groups. The explanation and fix for this problem was detailed in Chapter 8 of *Golfonomics*, and even supplied as a free download from the publisher's website at www.wspc.com, but, as evidenced by the continuing problem of slow play, has not been broadly understood. One of the main problems with overcoming the problem of slow play due to golf course design is that it can manifest itself in a very counterintuitive fashion. The chapter explains how there are two related notions of rate of play, namely, the time it takes to play (hours per round) and the flow rate of play (golfers per hour). And perhaps unexpectedly, by slowing the time rate for some (more hours per round) it is possible to increase the time rate on average (that is, decrease the hours per round on average) and thus, increase the flow rate allowing for more rounds of golf in total. Chapter 2 explains this in detail while providing simple numerical examples of the phenomenon and suggesting possible lines of follow up research.

In Chapter 3, "Golf Course Waiting: The Good, and the Bad . . .," another aspect of waiting is examined, namely, the rationing by waiting model as applied to obtaining tee times on a crowded golf course. As introductory economics shows, when the price is set beneath the market clearing level, a situation of excess demand or shortage exists because the quantity demanded is greater than the quantity supplied. The shortage brings about a chain of events as the pressures to bring about an equilibrium get played out in dimensions other than a simple increase in price. Unfortunately for the market participants, most of these events are costly to consumers and potential consumers and wasteful to society as a whole. Time spent waiting in line, or dialing into an automated tee time reservation system, are only the tip of the iceberg. This chapter explains how pure queuing would work. In the real world, however, pure queuing is distorted by the many types of attempts to jump the queue or otherwise manipulate the system to one's advantage. The chapter explains how any type of "non-price rationing" is inefficient, favors some groups over others, and in this light, therefore, is not "fairer" than simply raising the price to the market clearing level. As just one example shows, middlemen who manipulate the reservation system for the Bethpage State Park golf courses in New York may be able to capture rents approaching four figures per round that would otherwise go to the taxpayers of the state. Other numerical examples also show how quickly the losses to taxpayers from inefficient pricing can add up.

I'm sure it is obvious that I am not a fan of subsidized, below market-clearing pricing, even for a good which I enthusiastically consume on a regular basis. It is not only the fact that I have to wait in line, or the fact that I feel bad about taking from some non-golfer taxpayer who is poorer than me. The problems outlined in Chapter 3 are bad enough, but the problem is potentially much worse as explained in Chapter 4, "Golf Course Waiting: . . . and the Ugly." The problem is, indeed, truly ugly in that the existence of subsidized municipal golf courses actually leads to an equilibrium with fewer rounds of golf played on fewer golf courses of overall lower quality. Yikes! How can this possibly be? The answer lies in the unintended consequences of intervening in markets by those with good intentions. In this case the good intention is to make golf more available and more affordable by subsidizing the price at a publicly-owned golf course. The unintended consequence is that builders of for-profit daily fee golf courses are placed at a disadvantage in competing against an entity that does not have to cover its costs. The daily fee golf course does not get built (or its construction gets delayed) and the municipal course with a more or less captive market can skimp on its maintenance budget to save money for the city or for the management company, and still be at capacity. In a simple model, instead of two courses splitting a larger number of rounds, the equilibrium has one course at capacity with lower quality selling a lower number of rounds in total. The self-serving gut reaction of those who gain from municipal golf course operations (that is, both golfers and middlemen who can manipulate the queue to their advantage) is to discount the possibility and avoid the cognitive dissonance by not thinking about it. However, strong statistical evidence is provided for the entry deterrence effect by looking at the establishment of 104 golf courses in the San Francisco Bay area over a more than 100-year period.

In Chapter 5, "Consistency or Heroics," the focus changes from golf course economics to the economics of professional golf. This chapter also has ramifications for employee compensation and bonus plans in the non-sporting economy. Owing to the highly nonlinear prize fund structure in golf tournaments, professional golfers can earn a lot of money by receiving occasional large paydays for superior performance even though their average performance over the course of the season may be only mediocre. Playing superbly only occasionally might even be better than a consistently good, but never great, performance profile. This brings up the question of which type of performance is preferred, not only in golf, but in any company's attempts to bring forth effort from its employees. Would the PGA TOUR rather have a different winner each week, depending upon who has the largest negative deviation from their average score, or a set of familiar names as leaders due to steady play that is consistently better than their peers? Would a firm rather have occasional flashes of brilliance from its employees or consistent steady performance? The analysis in Chapter 5 sheds light on these issues. By looking at the variance (and skewness) of individual performances over the course of a season, the chapter measures and compares the effects of increasing variance and skewness versus decreasing one's average score as the path to higher earnings. Although variance and skewness do play a significant roll in the earnings of professional golfers, the numbers show that consistent better than average performance by the top five golfers in 2002 (Tiger Woods, Phil Mickelson, Vijay Singh, Ernie Els, and Retief Goosen) is more important than the ability to occasionally go low by having a large variance or a large negative skewness in one's week to week performances.

Having shown that variance and skewness in one's distribution of scores is important, the question now arises of where do skewness and variance come from? This topic is addressed in Chapter 6, with its long-winded title, "Skills, Performance, and Earnings in the Tournament Compensation Model: Evidence from PGA TOUR Microdata." The microdata refers to the painstaking collection of week-by-week performance statistics to capture not only the average driving distance, for example, but also to capture the variance and skewness of this and the other skills. Thus, the extent to which having a good or bad week in scoring comes from a good or bad week in driving distance, or putting, or approach shots, etc. is explored. The chapter first adjusts the individual week-by-week data to account for the ease or difficulty of the tournament course. These adjustments improve the previous literature, by eliminating a bias caused by, for example, not adjusting driving distance for altitude. Further, a second type of improvement is possible. By looking at the week-to-week adjusted data, a distribution of adjusted skill measures for each golfer is developed, thus allowing the calculation of mean, variance, and skewness. The chapter builds, in step by step fashion, a full structural model of golfer earnings in which the distribution of skills leads to the distribution of scores which in turn leads to earnings. Along the way, simpler formulations are presented for comparison sake. The more complex formulations do outperform the simpler ones, but most of the basic inferences in this line of research, that skills like long straight drives, hitting greens, and making putts are important to performance and earnings, are confirmed.

Ready for some provocative gender politics? Chapter 7, "Gender Discrimination Revisited," reprises some analysis from *Golfonomics* and brings it up to date with new data. In *Golfonomics*, using 1998 data to compare performance and earnings on the PGA TOUR to that in the LPGA, I show how, controlling for skills, women were actually paid more than men by about 29% or $9,000 per tournament. That is, even though women make less than men in total, they actually make more than men on a per skill basis. This result can only be sustainable because men are excluded from the women's only events sponsored by the LPGA. And, except for a few cases, women self-segregate to the LPGA where their earnings potential is higher. Fast-forward to the 2008 data and the results are pretty much the same. Over the ten years from 1998 to 2008 both men and women professional golfer earnings have grown much faster than inflation. Women are still paid more than men relative to their exhibited skills, by roughly the same $9,000 per tournament, although the result is no longer statistically significant and now represents only about 9-12% of the total earnings on a per-tournament basis. The chapter describes how these conclusions are reached. The analysis is based on the quality statistics on golfer skills that allow the earnings gap between men and women to be decomposed into two parts–one attributable to discrimination, and one attributable to differences in skills or productivity. Such a decomposition is generally not possible in most areas of the economy because usable comparable data on multiple dimensions of productivity simply do not exist. Without a careful decomposition based on quality statistics, popular claims of gender inequity (of the sort that women earn only $0.78 for each dollar that men earn) amount to no more than inflammatory rhetoric.

Lest the reader get the wrong impression, I want to stress that I am a fan of the LPGA, and a fan of women golfers in general. I hoped to show in Chapter 8, "Gender and Driving Distance," that perhaps women could improve the level of their skills, especially driving distance where they faced the biggest gap with the men, and thus move closer to gender neutral competition in golf. I actually anxiously anticipate this day. To track this improvement, the chapter starts with a survey of more than a dozen studies of the golf production or earnings function spanning data from over 20 years. I specifically focused on driving distance because the differences between men and women on the other skills were much smaller. Driving distance was almost always an important determinant of scoring or earnings. Furthermore, the implied elasticity of earnings with respect to driving distance is in the elastic range in all but one study and averages over seven. The chapter moves to a regression model of the trends in the growth of driving distance over a recent 19-year span. Controlling for gender and age, driving distance for women appeared to be catching up to that for men from 1992 to 2002, but has since settled back to a typical deficit of about 38 yards. Unfortunately, without catching up in this skill, women's golf will remain at a disadvantage. The chapter concludes with an analysis of the relationship between the growth of prize funds and the growth of driving distance. Arguably, the causality could go either way–from larger prizes to longer drives through the increased incentive to practice, or from longer drives to larger prizes through the extra enjoyment and interest of golf fans. Furthermore, the causality could be different on the PGA TOUR from that in the LPGA, and the chapter shows how.

Chapter 9, "To Play or not to Play: The Skills Match," makes further use of the tournament-by-tournament microdata discussed in Chapter 6. It exploits the facts that different golfers display different amounts of each golf skill, and different tournaments reward each of the skills differentially. Golfers who choose to enter those tournaments that reward the mix of skills they possess, should have more success than golfers who do not. The end results of the analysis confirm this suspicion, especially so for about half of the golfers in the sample of the 100 top money winners from 2005. Along the way, several intermediate steps in the analysis are interesting in their own right. Looking at individual results of over 2000 entries into tournaments it is possible to quantify how much of each skill each golfer has, and how important each skill is at each tournament venue. Interestingly, putting and hitting greens in regulation are the first and second most important skills in 40 of the 46 tournaments studied, meanwhile, driving distance is at best the second most important, and this happens only once. With the skills in hand measured for each golfer and the skills needed measured for each tournament, the next step is to get a ranking of expected results for each golfer in each tournament. Using these rankings, golfers can determine in which tournaments they should expect the most success, either in terms of scoring or potential money earned, and the researcher can determine how well the golfers do in choosing. The problem is actually equivalent to the combinatorial problem of choosing numbered balls without replacement from an urn. Consider an urn with 46 balls numbered from 1-46. A golfer choosing to play in the best 23 tournaments for him personally is analogous to choosing balls numbered 1-23, and the probability of doing so randomly (less than one in a trillion) is a straightforward calculation using combinatorial arithmetic. Each golfer's choices are examined and the probabilities of randomly choosing so well are calculated. The null hypothesis of random tournament selection is rejected for about one half of the sample. Furthermore, the better the choices made by the golfers, the more money they earn per tournament, controlling for their individual skill levels. The chapter is path-breaking in the sense that there is no other similar analysis that bases its hypothesis testing on comparisons of actual choice to theoretical combinatorial distributions.

In Chapter 10, "To Enter or not to Enter: Another Look," we, indeed, take another look at the entry decision of professional golfers. The combinatorial analysis in Chapter 9 has its advantages, but it is only capable of looking at one ranking factor at a time. Economists use the more familiar method of multiple regression analysis to simultaneously consider many factors that can affect a decision to enter a tournament. The degree of match between the golfer's strengths and the tournament's requirements is one factor, but age, earnings, injury status, and dynamic factors are also important. Especially so are dynamic factors such as where a golfer stands on the yearly earnings list, because ending in the top 125 on the earnings list is of paramount importance to keep one's playing privileges for the next year. Ending the year in the top 30 is also highly desirable because it allows entry into special no-cut, large purse tournaments. The chapter presents these regressions and compares the results to the combinatorial analysis from Chapter 9. Even after considering these extra explanatory variables, the degree of match between the skills in hand and the skills required is still significant for about one half of the golfers in the sample.

In Chapter 11, "To Bet or not to Bet: Sports Gambling, Golf, and Efficient Markets," a new data set of posted odds for golfers in PGA TOUR events is examined to assess two types of market efficiency. Sports economists have often looked at gambling markets in horse racing and many team sports to determine whether profitable betting algorithms could be discovered. This chapter is the first to do so for professional golf. The informational efficiency of a gambling market is generally determined by whether the posted odds (or point spread) accurately predict the outcome of the competition better than any other available information at the time of the bet. If other information is also valuable, then the odds or point spread may be biased and a profitable opportunity may exist. The chapter fully explains the argument and develops numerical examples to show how competition works in such markets. Thus, the chapter is another example of sports statistics being used to illustrate economic market phenomena. Previous results showing market efficiency and the absence of profitable opportunities are reconfirmed here with the golf wagering data. But the chapter also looks at another type of market efficiency, having to do with the transactions costs involved in placing bets. The chapter explains a couple of reasons why wagering on golf is more expensive than point spread gambling in football or basketball. While it costs a little less than $1.05 to cover all positions and win $1.00 with certainty (disregarding ties) in football, the comparable cost in golf averaged $1.51 for the posted odds in PGA TOUR events in 2002.

Switching gears again, Chapter 12, "Still Looking for Economic Impact: The Case of the Golf Majors," looks at 36 Major golf tournaments at 24 different courses over 18 years to measure the amount of local economic activity that the tournaments bring to the counties in question. The short answer is . . . none. The chapter explains the difference between *ex ante* studies, which typically promise large and multiplied increases in economic activity from hosting mega events, and *ex post* studies, which use actual data after the fact and almost always show that studies of the former type are overly optimistic. The chapter explains how and why the *ex ante* studies are flawed. The how has to do with substitution effects, double counting, and flawed application of the multiplier model; the why has to do with politics, advocacy, and boosterism which distorts and abuses economics in pursuit of supporting rhetoric as opposed to a more objective economic analysis which would seek to make an accurate rather than a rosy prediction. The golf Majors with rotating venues afford the perfect opportunity for a controlled study that looks at multiple sites over multiple years with each county that hosts a Major serving as a member of the control group in the years that it does not host one. Unfortunately, there is no evidence of increased employment either in the year of hosting a Major or in several years afterwards.

Finally, as I prepare to hang up the chalk (but not the spikes) it gratifies me that several other researchers have started to mine topics in golf economics and data from golf to inform other economic theories. In Chapter 13, "The Best of the Rest," I briefly summarize the direction that other economists have taken. I cannot cover them all and apologize in advance for several errors of omission, but the sample of research that I do cover gives an idea of the originality and breadth to which the economics of golf has grown.

My friend and colleague, Jagdish Agrawal looks at the value to companies of celebrity endorsements. And there is, perhaps, no one with more stature than Tiger Woods in this regard. Tiger's case is made even more interesting by the ups and downs of his unfortunate personal behavior, and he [Agrawal and Kamakura, 1995; Agrawal *et al.*, 2013] uses the event study methodology to ferret out the effectiveness of Tiger and others as celebrity endorsers.

Recognizing that sports promoters want to design contests and prize structures to bring forth great displays of talent and effort for the fans to enjoy, Jennifer Brown tackles the thorny issue of effort provision in the contest design literature.[3] Simple economics suggests that effort provision should be subject to the same type of cost-benefit analysis as any other decision. Brown notes how there might be subtle differences in the cost-benefit calculus coming from competing against a "superstar." A potential problem occurs if one contestant, the superstar, is superior to the others in such a way that he or she can win simply by going through the motions, and such that the other competitors don't try hard either. Again it is Tiger Woods that supplies the setting for a test of these propositions. By looking at over 30,000 different scores Brown compares how golfers other than Tiger perform in three different settings: When Tiger is not playing; when Tiger is playing during one of his hot, dominant spells; and when Tiger is

[3]See Brown, J. (2011) Quitters Never Win: the (Adverse) Incentive Effects of Competing with Superstars, *Journal of Political Economy*, 119(5), pp. 982-1013.

playing during one of his (relatively) cooler periods. The most difficult part of the research is controlling for the difficulty of the golf course. Brown cleverly handles the problem by exploiting changes in Tiger's schedule from year to year to achieve measures of golf course difficulty independent of whether Tiger is playing.

Seung Chan Ahn and Young Hoon Lee combine the issues of subtle differences in incentives on the one hand and potential earnings from celebrity endorsements on the other in their study relating beauty to golfing performance.[4] One might ask why should beauty have anything to do with playing exceptional, championship golf? The link is that better looking people may be able to earn more as celebrity endorsers than plain or homely people. If so, the potential returns to winning an important golf tournament are higher for the better looking. Going backwards one step, the potential benefits to developing golf skill are higher for the better looking and, as such, the better looking have more incentive to practice long and hard. Using careful but subjective surveys of beauty on the LPGA circuit, Ahn and Lee are able to tease out a significant nonlinear relationship between beauty and playing well that confirms their theory and has no other obvious explanation. The research holds interesting implications for psychology and sociology as well as for the economics of discrimination.

Economists Devin G. Pope and Maurice E. Schweitzer have been able to exploit a large data set with information about 2.5 million putts hit by PGA TOUR professionals to identify and confirm a prediction from a relatively new field called behavioral economics.[5] The specific prediction involves "loss aversion" which asserts that something that is had and lost is worth more than if it was never had in the first place. Thus, professional golfers, who may consider already having a par to be the usual outcome on a hole, will think of missing the par putt as losing the par. This will hurt more than missing a birdie putt which may be considered a lost opportunity but not a loss of something that should have been had. In this way, the par

[4]See Ahn, S. C. and Lee, Y. H. (2014) Beauty and Productivity: The Case of the Ladies Professional Golf Association, *Contemporary Economic Policy*, 32(1), pp. 155-168.

[5]See, Pope, D. G. and Schweitzer, M. E. (2011) Is Tiger Woods Loss Averse? Persistent Bias in the Face of Experience, Competition, and High Stakes, *American Economic Review*, 101(1) pp. 129-57.

putt seems more important than the birdie putt, even though a stroke is a stroke is a stroke, and thus the golfer will concentrate more and make more par putts than birdie putts of similar distances. Insights from behavioral economics have been verified in experimental settings but Pope and Schweitzer's work is important because the usual criticisms of low stakes and unmotivated or uninformed decision makers do not apply to professional golfers on the PGA TOUR.

The final section of the final chapter provides a springboard for new research using the new golf statistics. When I started doing golf research in the 1980's all that was available was year end averages in several statistical categories. I developed my own tournament-by-tournament data at great effort but this was limited to one year and led to the material in Chapters 5, 6, 9, and 10. Now, using laser measuring devices and volunteers, data exists for every shot in every tournament, and the PGA TOUR will allow access to it for the purposes of academic research. Thus, the frequency of observation in the available data has increased from yearly, to tournament, to individual shot over the last three decades. In the new data, each shot can be assigned a stroke value based upon the ball's starting and finishing points and the average performance of the rest of the field from these same spots. The chapter explains the new method and runs a simple model showing how the new statistics are superior to the old ones. At least that is what it was supposed to show. In the simple model the older statistics actually work slightly better. I consider this to be a puzzle, and in the book I offer a prize for the best explanation. You'll have to read the book to find out more.

Chapter 2

The Economics of Slow Play[1]

When a three and one-half hour round of golf stretches into five or six hours, it's bad for the golfers and bad for the golf course owners. Virtually all golfers and golf course managers know that it is bad, but have been largely unsuccessful in attempts to speed the pace of play. Earlier efforts by the United States Golf Association (USGA) introduced in the *1995 USGA Pace Rating System*[2] targeted things that golf course managers could do to avoid long rounds. But these have not rid the game of the problem of slow play. Although the USGA's pace rating system makes some strides in the right direction, and offers some good suggestions, it is hampered by an incomplete understanding of the causes of slow play and of the most efficacious ways of dealing with them.

More recently, the USGA and the PGA of America have introduced new efforts that target golfer behavior. The "Tee it Forward" campaign, with cameos by Jack Nicklaus, urges golfers to play faster by playing from shorter tees. Presumably the gain in speed comes from taking fewer strokes. I might sardonically suggest a "Hit it Straighter" campaign to achieve the same purpose. Skipping the even numbered holes would also

[1]This chapter is a repeat of Chapter 8 from *Golfonomics*. The material in it is important enough to be in both books and, unfortunately, the advice in it has heretofore gone unheeded. A earlier version of this chapter was presented at the Western Economic Association, International meetings in San Francisco, July, 2001.

[2]*1995 USGA Pace Rating System*, (United States Golf Association, Far Hills, New Jersey), 1995.

work. Not to be outdone, during 2013 the USGA unveiled its "While We're Young" campaign featuring Arnold Palmer. Using light humor, the campaign urges golfers to not foolishly waste time like the golfer in the commercial who stops in the middle of addressing the ball to go to the beverage cart. If it was the case that such capriciously silly and rude behavior was the major cause of slow play, then the problem could be easily solved this way. But it is not. The "While We're Young" campaign provided a few chuckles, but does anyone really think that the problem of lengthy rounds of golf can be solved so simply? Indeed, industry expert Bill Yates has identified player ability and player behavior as only secondary and tertiary causes of slow play.[3] Thus, "Tee it Forward" and "While We're Young" will also fail because they do not address the main problem.

This chapter extends our understanding of slow play by employing some insights from production theory, specifically, the attempts to solve production bottlenecks, as developed in economics and management science. An analogy is drawn between the flow of golfers around a golf course and the flow of partially finished inventory through a factory, explaining how the idea that is variously called Just-In-Time (JIT) production, Zero Inventory Policy (ZIP), or production smoothing, can be applied to help move the maximum number of golfers most comfortably around a golf course.

The remainder of this chapter proceeds as follows. A brief introduction to the relevant issues in production theory is presented first. Following this, the similarities between golfers on a golf course and inventory in a factory are developed. Once the analogy is developed, complicating factors, such as the difference between the rate capacity and the volume capacity, will be introduced. This difference, which is not fully appreciated in the *USGA Pace Rating System*, is of crucial importance to understanding and solving the bottleneck problem. After these preliminaries, the bottleneck problem itself is introduced and discussed fully. To further illustrate the bottleneck problem, the chapter develops a stylized, abstract model, comparing three common causes of delays on a golf course and forming the basis for future empirical study. Finally, the abstract model is

[3]See, Newport, J. P. (2013) The Real Causes of Slow Play, *Wall Street Journal*, July 13-14.

applied to the bottleneck problem on a par three hole to illustrate how the issues of bottlenecks, production smoothing, and rate and volume capacities interact to produce the optimal flow of golfers around a golf course.

2.1 Just-In-Time Production

A production or assembly plant is typically composed of numerous stations where workers or machines sequentially perform the many discrete steps necessary to produce the product. If the first step can be completed more quickly than the second step, then an inventory will build up. For example, suppose the first step takes 15 minutes (alternatively called a production rate of four per hour) and the second step takes 30 minutes (two per hour). After an hour of production, the station performing the first step has produced a batch of four items, two of which have moved through to completion of the second step, and two of which form an inventory of items finished with step one, but waiting for step two.[4] For the next hour, the station performing the first step can rest, or be employed with another task, while the station performing the second step uses up the inventory. Call this the batch system of production.

There are both costs and benefits of employing a production system like the one described above. The obvious benefit of building up an inventory of items that have completed the first station is that it can be closed for scheduled or unscheduled repairs or maintenance, or employee absence, without closing the production line. Station two can continue to operate using the inventory that has been built up. The obvious cost of such a system is that money is tied up in partially completed inventory. A less obvious cost is that mistakes being made at the first station may not be caught immediately. A whole batch containing mistakes could be made, temporarily stored in inventory, and eventually moved to the second station, where only then is it discovered that an unacceptable percentage of the batch does not measure within the required tolerances. Perhaps the

[4]This is not precisely correct if there is no inventory to start with for the second step. In this case the second step is idle for fifteen minutes awaiting the completion of the first unit in the first step.

faulty items could be fixed, adding a cost, or, in the worst case, the items may prove to be pure waste.

In what is now a familiar story to economists and production engineers, the "Japanese-style" production processes as pioneered by Toyota, involve a relentless attempt to reduce inventories throughout the factory. The result is known as production smoothing, ZIP, or JIT. Along a production line, wherever an inventory is seen to build up, that is the signal to employees and managers that the step after the inventory is too slow; it is a "bottleneck" in the production line which is slowing down the rest of the plant. In the example above, station two is the bottleneck. Recognizing the bottleneck, managers know that attempts to speed the completion of step one will have no effect except to build up bigger inventories of possibly defective items. If anything, step one should be slowed down until the items are being finished "just-in-time" for the second station to use them. This would reduce the inventory buildup but do nothing to increase the overall speed. All managerial efforts to increase speed should be targeted at the second station which is the cause of the bottleneck. Once production is "smoothed" between any two adjacent steps, a bottleneck may appear elsewhere along the production line. Management then directs its efforts to obtaining JIT or ZIP at that point in the factory.

Relative to the batch system, the JIT system has the obvious advantage of having less money tied up in inventory, and the less obvious advantage of avoiding waste by catching mistakes sooner. Furthermore, the JIT system has the advantage of fostering continuous improvements in the production process by highlighting bottlenecks and waste as opposed to hiding them with the use of inventories. However, the move from the batch system to a JIT system is not all gain. An extra cost of the JIT system is that whenever an unscheduled breakdown occurs, there is no inventory to tide the factory over until repairs can be made. Because of this, in the Japanese-style plant, the speed of repairs to production machinery is of utmost importance. Japanese workers even practice such speed by having competitive "races" between shifts to see which "team" can more quickly dismantle, troubleshoot, repair, and reassemble the broken machinery. By contrast, maintenance speed is of no importance if a batch of inventory exists.

2.2 JIT on the Golf Course

This very brief overview of JIT, ZIP, and production smoothing contains the insights we need to examine the pace of play on the golf course. The analogy proceeds as follows. The groups of golfers are the units of partially completed inventory. The golf course can be divided into multiple "stations" where the "production" takes place. The stations are the parts of each golf hole where only one group can play at a time. The production is the finishing of the round of golf by successfully completing each station, that is, each part of each hole.

For example, if the first hole is a 400 yard par four, it could be thought of as two stations. The first 250 yards of the hole is the first station. The group of golfers must hit their tee shots, travel to their balls, hit their approach shots and move clear of the first 250 yards at which time their production at the first station is finished and the next group, the second unit of production, can start. The first group then moves to the second station, that is, the last 150 yards of the hole where they must travel to the green, hit their chips, hole their putts and move clear of the green to the next station, the second tee. While the first group is on the green (the second station) the second group is moving up the fairway, that is, moving through the first station.

If the first station takes seven minutes to complete, and the second station takes seven minutes to complete, then the production is "smoothed." The first hole can be completed in 14 minutes. Tee times can be spaced seven minutes apart, and the groups will move through the hole, finishing seven minutes apart. Moreover, if the production is smoothed, the second group finishes the first part of the hole "just-in-time" to start the second part of the hole.

One measure of capacity in a factory is the rate of production, that is, how many units can be produced per unit of time. The rate capacity of the first hole is one group every seven minutes or 8.57 groups per hour. If there are four golfers per group, the rate capacity is about 34.3 golfers per hour. In addition to the rate of production, the total volume of production depends on the total time devoted to production. On a golf course the time is essentially the daylight hours. If a group wanted to finish just the first hole by dusk, they would have to start the first hole no later than 14 minutes before dusk. If there are 18 holes each taking 14 minutes, then the available time to start a round of golf is from daybreak to four hours and 12 minutes (4.2 hours) before dusk. To put this in perspective, consider the equinox and assume that there are exactly 12 hours between dawn and dusk. There are 7.8 useable hours for starting times on such a day. Multiplying this by the estimate of 34.3 golfers per hour, and rounding, yields the volume capacity of the golf course as 267 golfers per day. Obviously, the volume will be higher during the longer summer days and lower in the winter.

We can also talk about inventory, but we must extend the example to a second hole. An inventory of more than one group can never build up in the middle of the first hole because any succeeding groups cannot start the hole until the group in front of them has moved on to the second station, that is, moved up the fairway toward the green. However, groups of golfers can pile up and wait on succeeding tees. Therefore, it is possible to have an inventory of more than one group of golfers who have finished the second station, that is, putted out on the first green, but cannot yet start the third station, that is, hit their tee shots on the second hole.

2.3 Complications

The preceding distinction between where inventories can or cannot pile up on a golf course actually points to a nettlesome complication in pursuing the analogy between the flow of golfers around a golf course and the flow of inventory through a factory. The golf course example points out that some stations in a production process may be linked in a more fundamental way than simply being next to each other in sequence. In particular, it is possible for a group of golfers to be in two stations at once.

Consider the group of golfers in the fairway of the first hole waiting to hit their second shots. They are not yet in the second station because the group in front of them is on the green, so the group in the fairway has to wait. While waiting, they are still in the first station and the group on the tee cannot hit their tee shots. Once the group on the green is finished, the group in the fairway can play their approach shots which might take one or two minutes. During this interval, the group in the fairway is simultaneously in station two (using the green as the target for the shots) and in station one (using the part of the fairway that is the target for the tee shots of the next group).

For the example highlighted above, the following slight change in the numbers is necessary. Suppose it takes five minutes to hit tee shots, travel to the balls and line up (but not hit) the next shot. Suppose it takes two minutes to hit the approach shots, replace the divots, and clear away from the tee shot landing area. Finally, suppose it takes five more minutes to travel to the green, hit chips and putts, and clear away from the green. The group would be in station one for a total of seven minutes, and in station two for a total of seven minutes, and production would be smoothed with seven-minute intervals. But the hole takes only twelve minutes to play because for the two minutes required to hit the approach shots, the group is simultaneously in station one and in station two.

Because the first hole now takes only 12 minutes to play instead of 14 minutes, the volume capacity of the golf course increases, even while the rate capacity stays the same. Golfers will still be starting and finishing the first hole at seven minute intervals, so, as above, the rate of production is 34.3 golfers per hour. Because of the shorter time, however, there are now more tee times available within the daylight hours. For example, if the golf course was composed of 18 such holes, the total playing time would be 216 minutes or 3.6 hours. Tee times would then be available from dawn until 3.6 hours before dusk, for a total of 8.4 hours at the equinox. The volume capacity would increase to 288 golfers per day, the rounded product of 8.4 hours per day times 34.3 golfers per hour.

The important insight from this discussion is that the time it takes to play a hole and the rate at which golfers can finish a hole are two separate things. The rate is measured in golfers per hour, and the time is measured in minutes per hole. The two examples considered above have the same rates (that is, one group every seven minutes or 34.3 golfers per hour), but different times, 14 minutes versus 12 minutes.

A naive understanding of the relationship between these two measures can lead to an erroneous calculation. One might mistakenly reason that because par four holes can hold two groups at once, if the hole takes 14 minutes to play, then groups can be spaced at seven-minute intervals. This calculation works for the simple case but not for the case in which the two stations in the first hole overlap. Even though the first hole can be played in twelve minutes, groups cannot be started at six-minute intervals because the first fairway is not clear until the group ahead has played its approach shots.[5]

Unfortunately, it is clear from the *1995 USGA Pace Rating System* that the USGA has not completely understood this relationship. The USGA Pace Rating formula puts all its efforts into computing a "time par," which is the amount of time it should take for a group of four average golfers to complete a hole. The time par takes into account the length of the hole and the difficulty. The time par also factors in items such as restrictions of carts to cartpaths only, and long distances between a green and the next tee. The time par rating formula places no emphasis at all on the time it takes to play each station on each hole, or the time groups spend occupying more than one station. To its credit, the manual mentions in several places the need for correct spacing of golfers on a golf course and suggests ten-minute intervals between tee times. The ten-minute figure, however, is simply picked out of thin air. There is no attempt in the manual to link it to the numerous other time calculations that are made. This is doubly unfortunate for the golf industry because most golf courses use tee intervals that are closer than ten minutes. Many of these courses could benefit from some of the other good suggestions in the manual but management may be afraid to implement them because of the loss of revenue that extra spacing between tee times entails.

2.4 The Golf Course Bottleneck

By far, however, the most serious complicating factor that is absent so far harkens back to the production smoothing argument. In the above example, the seven-minute intervals could be maintained because the time it took to

[5]The calculation might be salvageable if one realizes that the group in the fairway is actually occupying two stations at once, so there cannot always be two groups on each par four hole. For two-sevenths of the time (two minutes out of every seven minutes) there can only be one group on the hole, so on average, there are (2(5/7) + 1(2/7) = 12/7) twelve-sevenths groups on the hole. Dividing twelve minutes per hole by twelve-sevenths groups per hole yields seven minutes per group.

complete each station was the same, seven minutes. In the first case, the seven minutes were not overlapping and, in the second case, the seven minutes overlapped by two minutes when the golfers were in the fairway hitting their approach shots. But JIT was established at seven-minute intervals.

Now, suppose that for a simple, non-overlapping case, the first station takes six minutes and the second station takes eight minutes. The hole still takes 14 minutes to play and, as such, is indistinguishable from our first example as far as time par and the USGA Pace Rating manual is concerned. Unfortunately, in this case, seven-minute tee time intervals will be disastrous. Suppose the first group tees off at 6:00 AM. At 6:06 AM the group would be clear of the landing area in the first fairway and the second group would be chomping at the bit to begin their round of golf. The starter, a paid or volunteer employee of the golf course who regulates when each successive group of golfers can tee off, will not allow the next group to tee off until 6:07 AM. Six minutes later, at 6:13 AM, the second group could have already hit their approach shots to the green if no one was in front of them. Unfortunately, the green would not have been clear until 6:14 AM, at which time: the first group finishes the second station; the second group begins the second station; and the third group (on the tee) begins the first station.[6]

So far, so good. The third group started right on schedule at 6:14 AM and could have played their approach shots and cleared the fairway by 6:20 AM except for the second group which would still be on the green until 6:22 AM, that is, eight minutes after the previous group has cleared the green. Now the problem starts to manifest itself. The fourth group, with a tee time of 6:21 AM, would find the first fairway not clear until 6:22 AM. Each succeeding group would find its tee time delayed by an additional minute.

Unfortunately, on an actual golf course this particular problem would be concealed by a variety of different factors. The first factor is that not all golfers can hit their tee shots 250 yards. For example, a group of ladies, beginners, or senior citizens might not have to wait for the fairway to be

[6]The simultaneous starts of the second station by the second group, and the first station by the third group, are due to the simplifying non-overlapping assumption.

clear for 250 yards before playing their tee shots. They could tee off on time with the group in front of them still waiting in the fairway. The group following them might also be able to tee off on time if the short-hitting group plays their second shots short of the green while the group in front of the short-hitting group is still on the green. Sooner or later, however, a group will arrive to tee off and find all the accumulated waiting staring them in the face.

A second factor that can disguise a first-tee bottleneck is when a group shows up to the first tee a minute or two late, and the golfers and the starter might not even notice the delay until the following group. By this time, the starting times might already be four or five minutes behind schedule, and the golfers and the starter would be gazing up the first fairway wondering which group is responsible for the slow play. If one group appears to be playing particularly poorly, or appears to be not ready to hit their shots because of searching for balls in the woods or high grass, then the temptation would be to blame the slow play on the slow group of golfers rather than the "unsmoothed" production timing of the stations on the first hole.

A third disguising factor is that a whole group may fail to show up for their tee time, at which point the golf course can get back on schedule. Note that no-shows are much more common on weekdays than on weekends, and it is on weekends that the problem of slow play is most acute. Also note that some golf courses make it a practice to leave one tee time open each hour. These golf courses will sometimes allow "walk-ons" (golfers without reservations), to tee off in this open spot if the course is not behind schedule. More often than not, the open tee time is skipped over to get the first tee back on schedule. Skipping the open tee time simply converts the seven-minute tee time intervals into realized tee time intervals that are approximately eight minutes apart.[7]

By far, however, the most important disguising factor is that the bottleneck described here might happen on a hole other than the first, say, the fifth hole, out of the direct view of the starters and managers, and after

[7]Seven-minute tee intervals translate into 8.57 (that is, 60 divided by 7) groups per hour. Consequently, if one time is unfilled, then 7.57 groups per hour translates into 7.93 (that is, 60 divided by 7.57) minutes between groups.

four holes of other random factors that influence the speed of play have occurred. In this case, all groups will have started on time but the backup will occur on the fifth tee. There are many courses that suffer this bottleneck problem somewhere on the golf course. Two or three groups can pile up on a tee and bemoan the slow play, but trade assurances with each other that, "it opens up after this hole." Of course it opens up. Golfers are arriving at the bottleneck station at roughly seven-minute intervals and finishing the bottleneck station at roughly eight-minute intervals. There is no avoiding the resulting backed-up "inventory" of golfers.

Any one of the above factors will disguise what is going on. Some groups seem to be on time and others have to wait long intervals. The first reaction of many golf course managers and marshals would be to bemoan the slow play of those groups directly in front of groups that seem to be waiting the longest. Because the early-morning groups are least affected by the cumulative nature of the backup, they can easily play their rounds within a decent time interval. Afternoon groups are those most slowed down. The fact that some groups meet the time targets while others cannot, reinforces, in the manager's mind, the conclusion that the problem lies with golfers not with the golf course. Thus, the "solution" is to educate and encourage golfers to play faster through a variety of "fast-play" tips, warnings, and incentives, including: keeping up with the group in front of you; playing out of turn (called, "ready golf"); raffles or lotteries for those meeting time targets; teeing it forward; doing it while we're young; and being forced to skip holes to catch up with the group in front of you if you fall behind. Although this last threat is rarely, if ever, carried out, golf course managers must think that it has some efficacy as a signal of the seriousness with which the golf course takes the slow-play problem.[8]

Even the USGA manual does not seem to appreciate the nature of the problem. In a question and answer section near the end of the manual the following question is posed. "How does the USGA Pace Rating address

[8]In this author's view it is unlikely that a golf course would expect its marshals (who are usually elderly, retired gentlemen volunteering their time in return for free golf) to take on the policing and enforcing role of telling a possibly surly, possibly inebriated, and possibly frustrated by their own poor performance, group of golfers to pick up their golf balls and skip the next hole.

the common experience that the first groups of the day play at an acceptable pace but later groups take longer and longer to finish their rounds?"[9] The answer given in the manual demonstrates that the USGA does not fully recognize the bottleneck problem even when the pattern of slow play that the bottleneck causes is directly observed. In fact, the answer starts by rephrasing the question into one that is not equivalent.

> A. The issue being raised is, "How can a group catch up after it has fallen behind?" First, an awareness of time par allows a group to know how fast it should be playing. Time par takes into account the length and difficulty of the hole and the number of shots it should take a group of four average golfers to reach the green. With time par, golfers have a benchmark time to shoot at on every hole. Because time par assumes a full course, it is a legitimate expectation for golfers on a busy day.[10]

The answer goes on to mention adequate spacing of starting times and sufficient maintenance on difficult holes but without any explanation of how they are connected to the Pace Rating, how they help a slow group catch up, or how they are related to the original question about fast morning rounds and increasingly slow rounds as the day progresses. The proper answer to the question of how the Pace Rating System addresses the longer and longer afternoon rounds is that it does not.

The problem does not lie with any one particular group. The problem lies with the golf course design. To use the language of production theory, production is not smoothed and a bottleneck exists. If tee times are spaced at seven-minute intervals, and if a particular part of the course that only one group can occupy takes eight minutes to play, there is no way around the conclusion that with each successive group, another minute will be lost.

Encouraging faster play by golfers in general would clear the bottleneck if the golfers could take one minute off the completion time for each station that they had to move through on the golf course. If what

[9] *1995 USGA Pace Rating System*, p. 42.

[10] *1995 USGA Pace Rating System*, p. 42.

would normally take six, seven, or eight minutes could take only five, six, or seven minutes, then seven-minute tee intervals could be maintained without a problem. The whole round of golf would also be completed in much less time. But, to avoid the bottleneck it is not necessary for the golfers to speed up on every hole. To avoid the bottleneck, the extra speed is needed only at that one step. It is unlikely, however, that golfers can systematically turn on the speed when it is required. Golfers' habits are developed over many rounds of golf and over eighteen holes in each round.

While encouraging a faster pace of play is always a good idea, an immediate fix for the bottleneck is to increase the tee time interval to eight minutes. Indeed, the Pace Rating system manual suggests longer tee time intervals as a way to consistently meet the time par, but it does not say why. Unfortunately, longer tee time intervals can appear to represent a direct loss in revenue to the golf course because they seem to imply fewer tee times each day. For this reason they may not be used.

The quibbling language in the preceding paragraph ("can appear to," "seem to," "may not be") is there for two reasons. First, the loss of tee times may not be real. Golfers will pay full price for a round of golf only if they expect to be able to finish 18 holes before dark. If longer tee-time intervals translate into shorter rounds, golf courses may be able to schedule full price tee times later in the day making up for fewer tee times all day long, and ending up with the same number of tee times. However, if starting-time intervals are too long, then fewer rounds would be scheduled and revenue would fall. The stylized examples later in the chapter will illustrate these effects. As previously mentioned, there is no systematic discussion of how to set the tee-time intervals in the USGA manual.

Second, with respect to the simple economics of this issue the statement that fewer tee times reduce revenue is strictly true only if price is constant. But price could go up for two reasons: (1) the rounds are faster and golfers will pay more because of the reduction in waiting time, and (2) market clearing price is higher when you slide back up the demand curve.

In any case, the better solution is to identify the bottleneck and eliminate it. Production smoothing can occur by slowing all the stations to the speed or capacity of the slowest station, and can offer efficiencies in the form of fewer inventories (in this case fewer golfers waiting in backups on the course) but it can also occur by concentrating efforts to speed up the slowest station, thereby increasing overall capacity. The example in the last part of the chapter, which examines the issue of calling up the next group on a par three hole, nicely illustrates the distinction.

2.5 Three Types of Slow Play

The discussion and explanation of the golf course bottleneck is not meant to deny that there are other causes of slow play and delays on the golf course. For the most part, everyone understands that if one particular group is consistently playing slower than everyone else, then the holes in front of the slow group will be open and the holes behind will be backed up. Everyone also understands that every golfer, including professionals, can hit wayward shots and have to take extra time to find the ball, assess penalties, take relief according to the rules, and finish the hole. A particular part on the golf course might take an average of seven minutes to finish, but because of the game's inherent variability, there will be a deviation around the seven minutes. Will a group that takes a little longer on one hole be able to make it up by playing the next hole a little faster than average? Unfortunately, the answer is no.

The following stylized model illustrates that there are three patterns of delay caused by three different causes: the golf course bottleneck; the slow group; and the variance in completion times. The three patterns set the stage for empirical application to discover which of the three is most important.

Consider the following stylized model of play over four stations. This can also be thought of as any consecutive four stations on an eighteen-hole golf course. Ordinarily, each station takes seven minutes to play, and there is no overlapping of the stations. Golfers arrive at the first station at seven-minute intervals, with the first group arriving at time zero. Each table below tracks the timing of each of four groups through the four stations.

Table 2.1 is the simple benchmark case. The rows refer to groups 1-4. The columns headed by 1-4 indicate the four stations, or holes, on this golf "course." For each group on each hole the table shows an ordered pair representing the time they start and finish each hole. Group 1 starts the first station at time zero, and finishes it at time seven. The stations do not overlap so that at time seven, group 1 simultaneously finishes hole 1 and starts hole 2, while group 2 starts hole 1. In column 4, the second number indicates the time that the indicated group finishes the fourth station. To highlight the total finishing time for all four groups, the last number for group 4 on hole 4 is in bold italics.

Table 2.1 The benchmark, even-flow case.

#	w	1	w	2	w	3	w	4	interval	total
1		(0, 7)		(7, 14)		(14, 21)		(21, 28)	-	28
2		(7, 14)		(14, 21)		(21, 28)		(28, 35)	7	28
3		(14, 21)		(21, 28)		(28, 35)		(35, 42)	7	28
4		(21, 28)		(28, 35)		(35, 42)		(42, ***49***)	7	28

In this case it takes 49 minutes to move the four groups through the four stations. The column labeled "interval" refers to the elapsed time since the preceding group finished. In this case, the groups are always seven minutes apart. The column labeled "total" is the total number of minutes (in this case, 28), it takes each group to play. Finally, the columns headed by w indicate any waiting that takes place on the golf course. In this case the play is "smoothed" and no one waits.

Now, compare Table 2.1 to Table 2.2 which represents what happens if group 2 is slow and takes eight minutes to play each hole. The first group is unaffected since they are ahead of the slow group. The slow group starts on time but falls a minute further behind on each hole. They finish 11 minutes behind the first group and take 32 minutes in total to play. The groups following the slowpokes wait an additional minute on each tee which lengthens their round of golf to 32 minutes also. It now takes 53 minutes for these four groups to play, and if there are only 49 minutes of

Table 2.2 Slow group # 2.

#	w	1	w	2	w	3	w	4	interval	total
1		(0, 7)		(7, 14)		(14, 21)		(21, 28)	-	28
2		(7, 15)		(15, 23)		(23, 31)		(31, 39)	11	32
3	1	(15, 22)	1	(23, 30)	1	(31, 38)	1	(39, 46)	7	32
4	1	(22, 29)	1	(30, 37)	1	(38, 45)	1	(46, ***53***)	7	32

daylight available, then the fourth group will not finish. This table highlights the pattern of slow play caused by a slow group. One group can affect all the groups behind it if the course is full.

Would eight-minute tee intervals help in this case? The answer is yes and no. Some of the waiting can be eliminated, but the fast groups would eventually catch the slow group. Furthermore, as Table 2.3 shows, the capacity of the golf course would be reduced even further.

Table 2.3 Slow group # 2 with 8-minute start intervals.

#	w	1	w	2	w	3	w	4	interval	total
1		(0, 7)		(7, 14)		(14, 21)		(21, 28)	-	28
2		(8, 16)		(16, 24)		(24, 32)		(32, 40)	12	32
3		(16, 23)	1	(24, 31)	1	(32, 39)	1	(40, 47)	7	31
4		(24, 31)		(31, 38)	1	(39, 46)	1	(47, ***54***)	7	30

In Table 2.3 no one has to wait on the first tee. But group 3 would wait a minute on every hole except the first, and group 4 would wait a minute on every hole once they "catch the pack." The waiting is reduced but the capacity of the course shrinks even more because it now takes 54 minutes to get all four groups through, and tee times can only be scheduled until 54 minutes before dusk instead of 53 minutes as in the previous case.

From the first three tables it is clear that a slow group causes a pattern of one long finishing interval after which the golfers are bunched together, finishing at regular intervals. It is usually easy to spot such a delay by noting any big gaps between groups on the golf course.

Now let us see what pattern develops if there is a bottleneck. Suppose that hole 2 takes eight minutes to play. Table 2.4 illustrates the outcome. The groups start out at seven-minute intervals, and end up finishing at eight-minute intervals. This is accomplished by the ever-increasing backup at the second hole bottleneck. Even spacing between finishing groups at an interval longer than the starting time interval, coupled with an increasingly long wait somewhere on the golf course, is the telltale pattern if the problem is a bottleneck. This pattern differs sharply from the pattern that develops when a slow group is on the course. Empirically exploiting this difference can help to determine which type of slow play problem is plaguing a particular golf course.

Table 2.4 Bottleneck hole # 2.

#	w	1	w	2	w	3	w	4	interval	total
1		(0, 7)		(7, 15)		(15, 22)		(22, 29)	-	29
2		(7, 14)	1	(15, 23)		(23, 30)		(30, 37)	8	30
3		(14, 21)	2	(23, 31)		(31, 38)		(38, 45)	8	31
4		(21, 28)	3	(31, 39)		(39, 46)		(46, ***53***)	8	32

In the case of a hole that takes longer to play, production smoothing by increasing the starting-time interval can eliminate all waiting as is shown in Table 2.5. Instead of waiting on the golf course, the waiting is included in each group's tee reservation. The bottleneck disappears, and production is smoothed at eight-minute intervals but it still takes 53 minutes for the four groups to finish. Thus, increasing the tee-time interval would not reduce capacity for the bottleneck problem, but it would reduce capacity if the problem is a slow group.

Table 2.5 Bottleneck hole # 2 with 8-minute start intervals.

#	w	1	w	2	w	3	w	4	interval	total
1		(0, 7)		(7, 15)		(15, 22)		(22, 29)	-	29
2		(8, 15)		(15, 23)		(23, 30)		(30, 37)	8	29
3		(16, 23)		(23, 31)		(31, 38)		(38, 45)	8	29
4		(24, 31)		(31, 39)		(39, 46)		(46, ***53***)	8	29

A completely different reason for golf course delays comes from the variability of the time it takes to play each hole, or each station on a hole. In Table 2.1, each group took exactly seven minutes to play each hole. Now consider a very simple type of variation in which the average time is still seven minutes, but sometimes it takes six minutes and sometimes it takes eight minutes. The variability could be by the hole or by the group. Furthermore, the eight-minute time could come before or after the six-minute time. While these variations would balance out if the golf course was not crowded, the tables below show that typically, the variations do not balance out. This, indeed, is a common result in queuing theory. The fast times tend to be wasted, while the slow times cause backups that are not overcome.

First, consider variability by the group. Group 2, instead of playing the four holes in times of (7, 7, 7, 7), plays them with either of the following two patterns, (7, 8, 6, 7) or (7, 6, 8, 7). Table 2.6 tracks the first of these two patterns. The first group is unaffected by the variable play of the second group. Also, group 2 makes up its lost time's and finishes in 28 minutes total. Each group afterwards, however, has to wait a minute before starting the second station. Group 3 falls behind by one minute due to no fault of its own. Group 4 is right on the heels of group 3, making it appear as if group 3 is the culprit for slow play when, in fact, it is not.

Table 2.6 Group # 2 variability with pattern (7, 8, 6, 7).

#	w	1	w	2	w	3	w	4	interval	total
1		(0, 7)		(7, 14)		(14, 21)		(21, 28)	-	28
2		(7, 14)		(14, 22)		(22, 28)		(28, 35)	7	28
3		(14, 21)	1	(22, 29)		(29, 36)		(36, 43)	8	29
4		(21, 28)	1	(29, 36)		(36, 43)		(43, ***50***)	7	29

Will this problem cease to exist if the fast time comes first? The answer is no, as is shown in Table 2.7. The fast time is wasted as group 2 has to wait to start the third station anyway. Meanwhile, group 2's slow time on the third station will cause every group after group 2 to wait a minute before starting the third station. The main difference between this and the preceding case is that the blameless group 3 will not look like the culprit.

Table 2.7 Group # 2 variability with pattern (7, 6, 8, 7).

#	w	1	w	2	w	3	w	4	interval	total
1		(0, 7)		(7, 14)		(14, 21)		(21, 28)	-	28
2		(7, 14)		(14, 20)	1	(21, 29)		(29, 36)	8	29
3		(14, 21)		(21, 28)	1	(29, 36)		(36, 43)	7	29
4		(21, 28)		(28, 35)	1	(36, 43)		(43, ***50***)	7	29

Similar conclusions come if the variability is by station or "hole" rather than by group. Table 2.8 illustrates the case where the second hole is played in eight minutes by group 2 and in six minutes by group 3. Group

2 will fall a minute behind group 1. Group 3 will have to wait a minute to start the second station, which it finishes quickly, only to have to wait again at the third station. Each succeeding group will also have to wait a minute at the beginning of the third station. Unfortunately for golf course managers who are trying to discover the causes of slow play, the problem is well-disguised. The problem is the variability in the playing time of the second hole but it will be manifested by consistent, additional waiting on the third tee. This is the exact counterpart to Table 2.6 where group 2's variability makes it look like group 3 cannot keep pace.

Table 2.8 Hole # 2 variability with pattern (7, 8, 6, 7).

#	w	1	w	2	w	3	w	4	interval	total
1		(0, 7)		(7, 14)		(14, 21)		(21, 28)	-	28
2		(7, 14)		(14, 22)		(22, 29)		(29, 36)	8	29
3		(14, 21)	1	(22, 28)	1	(29, 36)		(36, 43)	7	29
4		(21, 28)		(28, 35)	1	(36, 43)		(43, ***50***)	7	29

A delay problem still occurs if the quick playing time at station 2 comes first. Table 2.9 covers the case where the playing times for the second hole are (7, 6, 8, 7). Again there is a delay. Group 2 plays the second hole quickly only to have to wait on the third tee. Group 3 would not have to wait, but every group after it would have to wait a minute to start the second station.

Table 2.9 Hole # 2 variability with pattern (7, 6, 8, 7).

#	w	1	w	2	w	3	w	4	interval	total
1		(0, 7)		(7, 14)		(14, 21)		(21, 28)	-	28
2		(7, 14)		(14, 20)	1	(21, 28)		(28, 35)	7	28
3		(14, 21)		(21, 29)		(29, 36)		(36, 43)	8	29
4		(21, 28)	1	(29, 36)		(36, 43)		(43, ***50***)	7	29

With any of these last four patterns, eight-minute starting intervals would eliminate the waiting on the golf course. But, as above in the comparison of Tables 2.2 and 2.3, it would take longer to move the four groups through the four holes thus reducing the capacity of the golf course. With eight-minute starting intervals, in a sense, each successive group is arriving not "just-in-time," but rather, a minute late. Most groups play most stations in seven minutes, but they are spaced eight minutes apart. While this is a comfortable flow that can be maintained, more "production" takes place with seven-minute tee intervals because each successive group is ready just-in-time (or slightly ahead of time). The "cost" of this approach is the occasional "inventories" of waiting golfers that are allowed to build up.

We have now come full circle to the discussion of production theory that opened the chapter. If one "station" on a golf course or one batch of production (group of golfers) takes eight minutes to complete, (all of the time or part of the time) inventory backups will occur (golfers would have to wait) if starting times are seven minutes apart. Production can be smoothed with the ZIP of stretching starting time intervals to eight minutes. This, however, comes at the cost of lower capacity. If efforts could be taken to convert the bottleneck eight-minute station into something that will withstand a seven-minute per group flow rate, then JIT, ZIP, and smoothed production can occur at seven-minute intervals. Furthermore, as the following application illustrates, even if the total production time (what the USGA calls the pace rating or the time-par) is longer, the capacity of the golf course can be increased if the flow rate of production is made faster.

2.6 An Application: Calling up the Next Group on Par Threes

The final section of this chapter develops a stylized model of the pacing of play on a golf course that has a par three hole that takes eight minutes to play, set into a balanced sequence of stations each taking seven minutes. Four successive stations are illustrated, with the second station being the slow one. As such, it is like the example covered in Tables 2.4 and 2.5. We will track the first 15 groups of golfers but the pattern of what happens will be clear after the first five or six groups, and will continue after the 15th group. Tables 2.10 and 2.11 are the counterparts of Tables 2.4 and 2.5 extended to fifteen groups. Table 2.10 shows the increasing waiting times at the second hole. By the tenth group of the day there will be two groups on the second tee because group 10 arrives after 70 minutes and group 9 does not start the hole until 71 minutes have elapsed. The finishing-time spacing is an even eight minutes but each group takes one minute longer than the group in front of it. It takes a total of 141 minutes to move all 15 groups through this four-station sequence.

Table 2.10 Bottleneck hole # 2 with 7-minute start intervals.

#	w	1	w	2	w	3	w	4	interval	total
1		(0, 7)		(7, 15)		(15, 22)		(22, 29)	-	29
2		(7, 14)	1	(15, 23)		(23, 30)		(30, 37)	8	30
3		(14, 21)	2	(23, 31)		(31, 38)		(38, 45)	8	31
4		(21, 28)	3	(31, 39)		(39, 46)		(46, 53)	8	32
5		(28, 35)	4	(39, 47)		(47, 54)		(54, 61)	8	33
6		(35, 42)	5	(47, 55)		(55, 62)		(62, 69)	8	34
7		(42, 49)	6	(55, 63)		(63, 70)		(70, 77)	8	35
8		(49, 56)	7	(63, 71)		(71, 78)		(78, 85)	8	36
9		(56, 63)	8	(71, 79)		(79, 86)		(86, 93)	8	37
10		(63, 70)	9	(79, 87)		(87, 94)		(94, 101)	8	38
11		(70, 77)	10	(87, 95)		(95, 102)		(102 ,109)	8	39
12		(77, 84)	11	(95, 103)		(103, 110)		(110, 117)	8	40
13		(84, 91)	12	(103, 111)		(111, 118)		(118, 125)	8	41
14		(91, 98)	13	(111, 119)		(119, 126)		(126, 133)	8	42
15		(98, 105)	14	(119, 127)		(127, 134)		(134, ***141***)	8	43

Table 2.11 Bottleneck hole # 2 with 8-minute start intervals.

#	w	1	w	2	w	3	w	4	interval	total
1		(0, 7)		(7, 15)		(15, 22)		(22, 29)	-	29
2		(8, 15)		(15, 23)		(23, 30)		(30, 37)	8	29
3		(16, 23)		(23, 31)		(31, 38)		(38, 45)	8	29
4		(24, 31)		(31, 39)		(39, 46)		(46, *53*)	8	29
5		(32, 39)		(39, 47)		(47, 54)		(54, 61)	8	29
6		(40, 47)		(47, 55)		(55, 62)		(62, 69)	8	29
7		(48, 55)		(55, 63)		(63, 70)		(70, 77)	8	29
8		(56, 63)		(63, 71)		(71, 78)		(78, 85)	8	29
9		(64, 71)		(71, 79)		(79, 86)		(86, 93)	8	29
10		(72, 79)		(79, 87)		(87, 94)		(94, 101)	8	29
11		(80, 87)		(87, 95)		(95, 102)		(102, 109)	8	29
12		(88, 95)		(95, 103)		(103, 110)		(110, 117)	8	29
13		(96, 103)		(103, 111)		(111, 118)		(118, 125)	8	29
14		(104, 111)		(111, 119)		(119, 126)		(126, 133)	8	29
15		(112, 119)		(119, 127)		(127, 134)		(134, ***141***)	8	29

Table 2.12 Bottleneck hole # 2 with 7-minute start intervals, call up next group.

#	w	1	w	2a	w	2b	w	2c	w
1		(0, 7)		(7, 10)		(10, 12)		(12, 13)	
2		(7, 14)	1	(15, 18)		(18, 20)		(20, 21)	3
3		(14, 21)		(21, 24)		(24, 26)		(26, 27)	
4		(21, 28)	1	(29, 32)		(32, 34)		(34, 35)	3
5		(28, 35)		(35, 38)		(38, 40)		(40, 41)	
6		(35, 42)	1	(43, 46)		(46, 48)		(48, 49)	3
7		(42, 49)		(49, 52)		(52, 54)		(54, 55)	
8		(49, 56)	1	(57, 60)		(60, 62)		(62, 63)	3
9		(56, 63)		(63, 66)		(66, 68)		(68, 69)	
10		(63, 70)	1	(71, 74)		(74, 76)		(76, 77)	3
11		(70, 77)		(77, 80)		(80, 82)		(82, 83)	
12		(77, 84)	1	(85, 88)		(88, 90)		(90, 91)	3
13		(84, 91)		(91, 94)		(94, 96)		(96, 97)	
14		(91, 98)	1	(99, 102)		(102, 104)		(104, 105)	3
15		(98, 105)		(105, 108)		(108, 110)		(110, 111)	

Table 2.11 shows what happens for eight-minute tee-time intervals. There is no waiting, groups finish at eight-minute intervals, but it still takes 141 minutes to move 15 groups through these four stations.

Now consider what happens if the golf course imposes the policy of calling up the following group on par three holes. Specifically, suppose that the eight minutes it takes to play this hole can be divided as follows: three minutes to hit tee shots toward the green; two minutes to walk to the hole; one minute to hit chip shots to the green; and two minutes to sink putts and clear the green. Traditionally, in between hitting the chip shots and putting, a group can stand aside and wait while the following group plays their tee shots to the green. Then, the first group putts while the following group walks toward the green. If the calling-up convention is not used, there is only one group on the hole at a time, and the green is idle for two minutes out of every eight minutes. By calling people up to the green before putting, there can actually be two groups on the hole for a portion of the time, and the green would be idle for only one out of each seven minutes on average as is verified in the following table. The hole can actually be divided into four stations, labeled 2a, 2b, 2c, and 2d. The pattern of play that develops is illustrated in Table 2.12.

Table 2.12 (*Continued*)

2d	w	3	w	4	interval	total
(13, 15)		(15, 22)		(22, 29)	-	29
(24, 26)		(26, 33)		(33, 40)	11	33
(27, 29)	4	(33, 40)		(40, 47)	7	33
(38, 40)		(40, 47)		(47, 54)	7	33
(41, 43)	4	(47, 54)		(54, 61)	7	33
(52, 54)		(54, 61)		(61, 68)	7	33
(55, 57)	4	(61, 68)		(68, 75)	7	33
(66, 68)		(68, 75)		(75, 82)	7	33
(69, 71)	4	(75, 82)		(82, 89)	7	33
(80, 82)		(82, 89)		(89, 96)	7	33
(83, 85)	4	(89, 96)		(96, 103)	7	33
(94, 96)		(96, 103)		(103, 110)	7	33
(97, 99)	4	(103, 110)		(110, 117)	7	33
(108, 110)		(110, 117)		(117, 124)	7	33
(111, 113)	4	(117, 124)		(124, ***131***)	7	33

The first group of the day is not affected by the calling-up policy. After they play their tee shots on the second hole (station 2a), walk toward the green (station 2b), and play their chip shots to the green (station 2c) only 13 minutes have elapsed since they started. The second group would not be finished with the first station until time 14 so the first group does not wait to call up the second group, since the second group is not ready. The first group takes two more minutes to putt out (station 2d), the second of which actually delays the second group, and then moves to station 3.

When group 2 comes to station 2, they find that the green is occupied and they have to wait for one minute until time 15. The second group members then play tee shots (time 15 to 18 in station 2a), walk toward the green (time 18 to 20 in station 2b), and play their chip shots to the green (time 20 to 21 in station 2c). At this point they turn back to the tee to see the third group ready to play at time 21. Group 2 then stands aside for three minutes while the third group members aim their tee shots toward the green.

The time is now 24 minutes after the starting time and simultaneously, group 2 is putting on the green (time 24 to 26 in station 2d) while group 3 is walking toward the green (time 24 to 26 in station 2b). It is this simultaneity that allows the gains from this policy. After putting out on the second hole the second group moves through the rest of the course, and finishes 11 minutes after the first group. Group 2 fell behind by four minutes because they waited one minute on the second tee and they waited three minutes on the second green while they called up the group behind them.

Meanwhile, the third group continues to play the second hole. They are ready to hit their chips just as group 2 vacates the green at time 26. This takes one minute. They then see that group 4 is not yet ready to hit their tee shots so there is no one to call up to the green. Group 3 putts out (time 27 to 29) and arrives at the third station only to wait four minutes for group 2 to finish the third station.[11] Meanwhile group 4 arrives at the second tee while group 3 is still putting and has to wait one minute for the green to clear.

After this, the slightly awkward pattern repeats itself with each group waiting a total of four minutes. Some groups have to wait a minute on the second tee and three more minutes on the second green while they call up the group behind them. Other groups are unimpeded on the second hole but have to wait four minutes on the third tee. But from the second group on, there is always a group ready to start the third station at regular seven-minute intervals. Therefore, production is smoothed and the seven-minute starting intervals are sustainable.

Standing aside and waiting makes playing the golf course take longer than in the previous case with eight-minute starting intervals. Except for the first group, each group takes 33 minutes to play the course, 29 minutes of playing time and 4 minutes of waiting time. But the 33 minutes can be reached with a starting rate of one group every seven minutes. In Table 2.11, it takes only the actual playing time of 29 minutes to play the course. There is no waiting, but groups can only start at eight-minute intervals. Overall, the calling-up policy increases the capacity of the golf course as it only takes 131 minutes to move 15 groups through these four stations. In Table 2.11 it takes 141 minutes to do the same job.

[11]It does not matter whether this group fails to call up the following group. If they do not call up the next group, they end up waiting on the next tee for four minutes as shown in the table. If they do call up the next group, it also slows them by four minutes: one minute of waiting for the next group to arrive on the tee; and three minutes while the next group hits their shots. In this case, they arrive on the third tee just-in-time to tee off.

Undoubtedly, this result will be somewhat counterintuitive to the golfers themselves. Waiting by the green to call up the group behind you seems to be an imposition that slows down your group, and delays the ultimate time when you will finish. How could it actually speed up play overall? Yet it does. It works by smoothing production over the eight-minute bottleneck station essentially by breaking the one eight-minute station into two or more stations that, although they take 12 minutes in total, they can be completed in seven-minute intervals without causing a bottleneck.

In addition to its counterintuitive nature, the calling-up policy can be difficult to implement. Consider group 2 in Table 2.12. They could finish three minutes faster if they did not call up the group behind them. Furthermore, being the first group to call up a following group, it appears as if they are a slow-playing group that cannot keep pace with those in front of them. The group in front of them finishes in 29 minutes while group 2 and all those who follow take 33 minutes. This is a pattern that is similar to the slow group pattern from Table 2.2 or Table 2.3. If there are sanctions for slow play or for not keeping up with the group in front of you, golfers would seem to be justified in being worried about the time it takes to call up the following group.

It is precisely this misunderstanding about the golf course bottleneck as a source of slow play that this chapter is trying to clear up. The managers at some golf courses focus only on gaps between groups. These managers quote silly cliches like, "keeping up with the group in front of you instead of keeping ahead of the group behind you." It is fine to promote fast play tips and ready-golf, but a course may be missing a chance to smooth production through a bottleneck par three hole by, for example, insisting on a calling-up policy that, to the naive, seems only to lengthen the round of golf. Many golf courses have signs at par three holes requiring groups to call up the following group. Unfortunately, it is not often generally understood how or why such a policy works, therefore, the signs go unheeded and the policy goes unenforced.

Other golf courses focus on the time par or pace rating as currently explained in the USGA manual. These courses are not in the position to recognize and smooth bottlenecks that occur on portions of a hole, because the time par is calculated on a hole-by-hole basis and not on a station-by-station basis. Golf courses can choose a lengthened interval between starting times. This reduces waiting and allows a round of golf to be played in a shorter amount of time, but the course may end up with a lower capacity than need be. There is no question that longer starting time intervals will work to reduce waiting on the golf course. The stylized models in this chapter clearly show why. But choosing this path is essentially the equivalent of slowing the production rate of all the machines in the factory to match the speed of the slowest one. By contrast, the insights from economic production theory, especially JIT and ZIP, encourage managers to find the slowest machine and speed it up. On the golf course this means finding the bottleneck and taking efforts to smooth play through it.

Chapter 3

Golf Course Waiting: The Good, the Bad . . .

The chance to play a finely manicured, challenging golf course that is also the venue for the United States Open golf championship for under $100, that seems like it is good, doesn't it? While most US Opens are held at exclusive private golf courses, and some are held at privately-owned resort courses with very high fees, the 2002 and 2009 US Opens were held at Bethpage Black, a course owned and operated by the New York State Park system, where current green fees are in line with many other municipal or publicly-owned golf courses throughout the country. Typically, municipal golf courses use public subsidies to keep the price low especially for residents of the state, county, or municipality in question. Sometimes the subsidies are explicit in that the golf course receives funds from the relevant jurisdiction. But even if this is not the case, and the golf course operations cover all the explicit costs, there are often subsidies in the opportunity cost sense as the golf course property could generate more revenue for the community's use than the low price policy allows. Simply stated, municipal golf courses allow some golfers to play for a lower monetary cost than otherwise and this is a benefit to those golfers.

Unfortunately, the benefit of lowered prices stops there and, as such, this chapter is misnamed in giving equal stature to the good and the bad. Simple economics shows how the bad outweighs the good. In fact, some of the bad is already hinted at in the above paragraph, namely, that the rest of the jurisdiction is made poorer by directly or indirectly subsidizing the golfers in question. While no one wants to focus on the losers in this financial arrangement, it is important to examine all the benefits and all the costs of subsidized price policies on publicly-owned golf courses. Armed with the complete information, most non-golfers and many golfers themselves would want to rethink local pricing policies.

This chapter develops a model of rationing by waiting that both highlights some of the hidden costs of below market clearing pricing and explains attempts by consumers to circumvent the queues that necessarily follow from such a policy. There are winners and losers from such pricing and the winners will be upset by the material in this chapter. Nevertheless, on balance it can be shown how and why the community is actually worse off. This chapter considers the good and the bad of subsidized prices, but it gets worse. The truly ugly is considered in the next chapter.

3.1. A Simple Model of Golf Course Demand and Cost

Figure 3.1 shows simple cost curves for a hypothetical golf course with a capacity of 100,000 rounds per year. This capacity is optimistic but not totally out of the question for a golf course that operates on a year-round basis. Consider a shotgun start with two foursomes on each tee. Eight times 18 yields 144 golfers. It is difficult but possible to stage two such events each day which yields over 100,000 (144 times 2 times 365 equals 105,120) rounds per year. Alternatively, a foursome every eight minutes yields 7.5 starting times per hour, which for 8.5 hours of starting times per day yields over 93,000 (4 times 7.5 times 8.5 times 365 equals 93,075) rounds per year. The 100,000 figure will make calculations easier.

The diagram shows a constant marginal cost per golfer of $5.00 per round out to the capacity limit. My previous research in *Golfonomics*[1] estimates the marginal cost per round at just under $3.00 per round using figures from 1994. With a little inflation the $5.00 figure seems reasonable for today. At 100,000 rounds the marginal cost becomes infinite, an obvious oversimplification which does not drive the results of the model.

[1]Shmanske, S. (2004a) *Golfonomics* (World Scientific Publishing Co., Inc., River Edge, NJ). See also Shmanske, S. (1999). The Economics of Golf Course Condition and Beauty, *Atlantic Economic Journal*, 27(3), pp. 301-313.

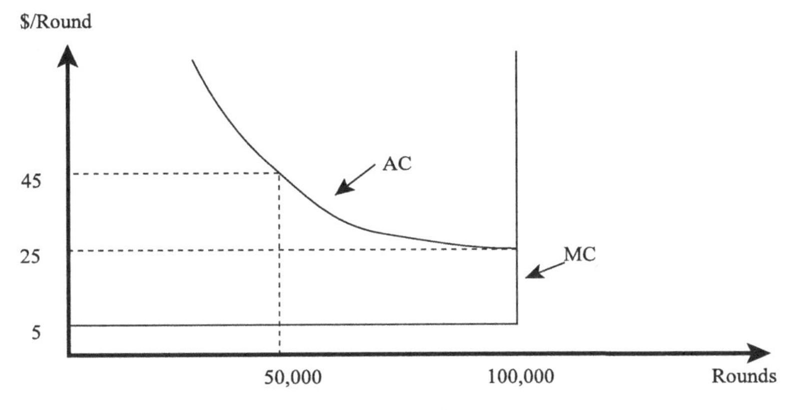

Fig. 3.1 Cost Curves.

Most of the costs of operating a golf course are overhead in nature and do not depend upon the number of rounds played. The green areas have to be watered, fertilized, aerated, and cut, the bunkers have to be maintained, and the pro shop has to be staffed even before one round is sold. The capital cost of debt service, or opportunity cost of invested capital are also independent of the number of rounds. If these overhead costs amount to $2,000,000 per year, then the average cost curve which adds the marginal cost to the overhead cost per round played is downward sloping throughout the relevant range. It bottoms out at the level of $25 per round at the capacity limit before also becoming vertical at that point. Note for future reference that the height of the average cost curve at 50,000 rounds is $45.00 and at 66,667 rounds the height is (rounded to) $35.00.

Now consider Fig. 3.2 which adds a demand curve to the picture. The market clearing price is $45.00 per round. The golf course could charge this price and return its excess cash flow of $2,000,000, or $20.00 per round, to the state or municipality for the benefit of all residents and taxpayers. However, as is typically the case, the publicly-owned golf course sells discounted rounds for the benefit of its resident golfers. In this model the golf course could cut price to $35.00 and still earn excess cash flow of $1,000,000. Indeed, the course could even lower price to the break even level of $25.00 per round without putting a strain on the community's resources except in the opportunity cost sense. In some cases, the municipal golf course even requires the infusion of funds from the government. If price were lowered to $20.00 per round, the golfers would be even happier, but the golf course would need $5.00 per round for a total of one half million dollars from the government to make ends meet.

Suppose the golf course pictured in Fig. 3.2 charges $25.00 per round. If so, there is a potential gain to the golfers who are able to obtain tee times without too much hassle. The gain is on the order of $20.00 per round, that is, the savings compared to the market clearing price of $45.00. It is this gain to these golfers that is the usual justification for the typical municipal, county, or state park system golf course operations. But, wait a minute, this

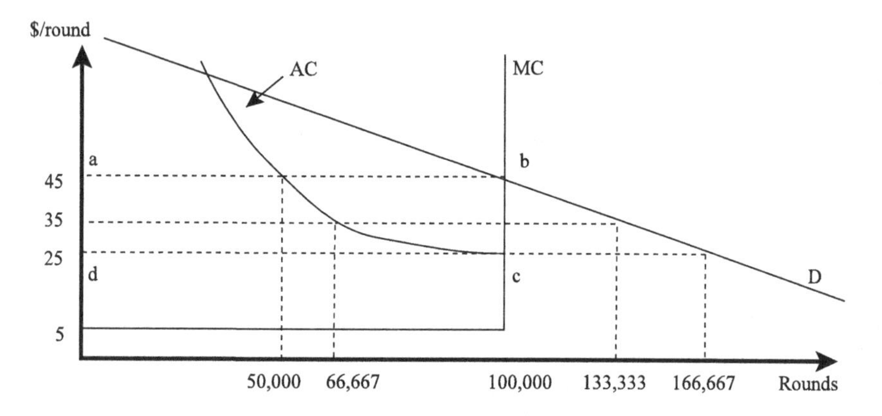

Fig. 3.2 Cost, Demand, Market Clearing Price, and Shortages.

gain is exactly offset by the $20.00 loss to the community's coffers. At best, the low price policy is a transfer of wealth from the rest of the community to the golfers. It is as if some taxpayer will be $20.00 poorer if the rest of the city's programs are to be maintained. Alternatively it is as if someone else's program has to be cut by the $20.00 per round transfer. The low price policy doesn't look so good anymore. Most golfers do not think about it, but if they did, they might become uncomfortable thinking that the money they saved on the green fee came from a fellow resident who might even be poorer or less well off than the golfer in question.

I note in passing that perhaps some golfers deny the existence of the subsidy because they do not understand the economic concept of opportunity cost. They can state with a straight face that there is no subsidy since no money goes from the public sector to the golf course. Of course, even this argument does not work if the price is lowered to the $20.00 level. But even if price is set at the $25.00 or $35.00 level, the denial of the resulting opportunity cost is a pitfall that the wise avoid.

Other golfers may recognize the subsidy but justify it to themselves by thinking that they deserve it because they too pay taxes, and that others may benefit from their taxes that go to pay for the community pool, the parks, and the schools which they themselves do not use. But this self-serving argument is not convincing. What about elderly taxpayers who cannot take advantage of any of these amenities? Wouldn't a better system be for the users of each amenity to pay their own way? Perhaps the community at large wants to subsidize schools or libraries for the positive spillovers that accrue to the rest of the community from having a literate, educated citizenry. But it is a stretch to think that this reason applies to subsidized golf. While it is true that golf courses are pretty, protect some habitat, and provide open space and other environmental benefits, these spillover benefits exist because of the existence of the golf course, not because of its subsidized price policies.

Perhaps the transfer of $20.00 per round from non-golfers to golfers (assuming a price of $25.00) could be tolerated if that were the only problem. Unfortunately, we are just getting started. Examination of Fig. 3.2 indicates that the quantity demanded at the price of $25.00 is 166,667 rounds per year. We now have the problem of distributing only 100,000 rounds to the 166,667 willing players. This is what economists call excess demand, or a shortage. Normally, the price does the rationing. In particular, if the golf course were trying to maximize profit it could raise the price to $45.00 (or higher) and the implicit subsidy as well as the excess demand for the 66,667 over-capacity rounds disappears. This is the beauty of the price-exchange system. But the low price policy does not allow the price to perform this function so some other method of rationing must be used.

3.2. A Simple Model of Rationing by Waiting[2]

Beginning students of economics will look at Fig. 3.2 and say there is a shortage at the price of $25.00. A slightly more advanced student might even say that there is a rent of $20.00 per round that is available for golfers and the rest of the community to split up as they see fit. However, the analysis should not end there because there are other implications. Mainly, since $25.00 is not an equilibrium price, there is pressure to increase the price. If the price is not allowed to rise, that pressure exerts itself in other, almost always counterproductive, ways. If the sellers set up no procedure to ration the demand, then the default procedure of queuing or lining up will naturally arise. In order to get one of the 100,000 rounds, you must be one of the first 100,000 in line. This means you have to beat out 66,667 others who also want to play.

At this point it is awkward to think of yearly demand and supply so I will change the numbers to daily rates. 100,000 rounds per year equals approximately 274 rounds per day.[3] The excess demand of 66,667 rounds per year translates to 182 rounds of excess demand per day, for a total demand per day of 456 rounds. These, of course, are averages. But for a golfer who wants to play about twice per week or about 100 rounds per year, it means that on each day of desired play the golfer has to beat out 182 others who also want to play but will not be able to. As an aside remember that this underlying demand curve is made up for illustrative purposes. Perhaps the excess demand is only ten percent of capacity, that is, 10,000 rounds per year. Now, on a desired day of play, the golfer has to beat out only thirteen or fourteen others, and this may change from day to day. Indeed, on rainy or otherwise undesirable days perhaps there is no excess demand, but on other days the excess demand might be for 30 or 40 tee times. This daily uncertainty about the level of excess demand is simply another hidden hardship that the golfers must endure.

So, getting back to the simple model, there will be a competition (think of it as a race) to be one of the first 274 in line for play on a particular day.

[2]For the original exposition of this argument see, Barzel, Y. (1974). A Theory of Rationing by Waiting, *Journal of Law and Economics*, 17, pp. 73-95.

[3]Some of the calculations in the previous chapter derive a figure of 267 golfers per day.

To see how this competition would play out, start from the premise that there was no waiting line. One could just show up when one wanted to play. This could lead to the prospect that all 456 golfers would show up at once and some prioritization method would have to single out 274 winners and 182 losers. Those losers would realize that by showing up one minute earlier they would move to the head of the line and get to play. However, others would endeavor to arrive two minutes earlier, others three minutes earlier, and so on. What is the equilibrium of this process?

The answer to the preceding question depends on the cost of arriving earlier and earlier. The simplest case occurs when everyone has the same cost of arriving early, say at a rate of $10.00 per hour. Now recall that a price of $45.00 will clear the market but that the golf course is charging only $25.00. If one has to show up two hours prior to the desired tee time, then the cost of waiting will be $20.00. This waiting cost added to the out of pocket cost of $25.00 brings the total cost up to the market clearing level of $45.00. That is, only 274 people per day will be willing to show up and wait two hours to play, and there will be enough tee times to accommodate them. The 182 people represented by points on the demand curve to the right of the capacity limit drop out because they are not willing to show up two hours in advance. Ordinarily consumers compete with each other to obtain a scarce good by offering or being willing to pay more money, but here, they compete by being willing to wait in line longer. The tragedy in this system is that the $20.00 per person of wasted time does no one any good, while a $20.00 increase in price goes directly into the public's coffers where it could reduce someone's taxes or increase the provision of some other needed public good or service. Over the course of a year the $20.00 per round of waiting cost amounts to two million dollars of wasted time instead of what could be two million dollars of lowered taxes or increased public services. This is represented by area abcd in Fig. 3.2.

3.3. Extensions of the Simple Model

In the simple equilibrium of rationing by waiting the available good ends up going to those who place the highest value on it. But this is a consequence of the simplifying assumption that everyone has the same value of time. Consider Fig. 3.3 which highlights two individuals in the points A and B. The person represented by point A is willing to pay $50.00 to play while the person represented by point B is only willing to pay $30.00 to play. If there was only one round of golf to distribute between A and B, most would say that societal benefit is maximized if A gets the round because he values it more highly. In fact, if B instead was entitled to the round of golf, then both A and B would gain if B could sell the round to A for any price between $50.00 and $30.00. For example, if B sold the round to A for $40.00, then A get something he values at $50.00 but only has to pay $40.00. Meanwhile, B gives up the round of golf which is worth $30.00 to him but gets $40.00 in return. Both are better off, and so is society in general, when the highest valued user consumes the good.

Now suppose that A and B have different costs of waiting. Person A's cost of waiting is $20.00 per hour and B's is $2.00. If the wait to play associated with a price of $25.00 is two hours, then A will not be willing to wait (his waiting cost of $40.00 added to the price is more than the $50.00 he is willing to pay) and B instead will be willing to wait in line (his

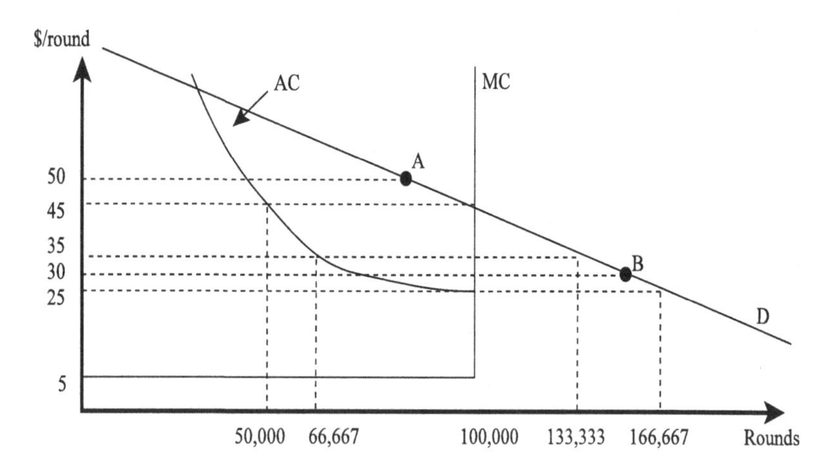

Fig. 3.3 Golfers with differing valuations.

waiting cost of $4.00 added to the price is still less than his willingness to pay). Thus, two hours of waiting corresponds to different dollar amounts of waiting cost and once differences in waiting costs are considered, it is no longer clear that the highest valued users are the ones who end up with the good.

People untrained in economics will sometimes claim that rationing by waiting is a fair method that gives each the opportunity to compete to get a scarce good for below the market clearing price. Put another way, some might claim that the highest value users should be determined by how long one was willing to wait in line. After all, the willingness to pay in dollar terms is influenced by one's overall wealth which is distributed unequally. Meanwhile, everyone gets the same 24 hours per day so that competing by waiting in line gives everyone the same chance.

A little deeper economic thinking shows that these claims are incorrect. First of all, it is not the case that everyone has the same ability to wait in line. In fact, each person's ability to wait differs depending upon, among other things, one's other commitments, one's age or health status, and one's physical proximity to the waiting queue. So, using queues instead of dollars simply replaces one unequal kind of ability and willingness to sacrifice with another unequal kind.

Second, using the willingness to sacrifice time by waiting in line to indicate intensity of demand does not translate into a meaningful value for the rest of society, while the willingness to sacrifice money does. No one gains by having people wait in line, or by having people exert themselves to learn how to wait in line better or faster. Alternatively, when money prices are used to ration demand, the resulting willingness to pay does represent real value. Indeed, obtaining the money in the first place represents the person's contributions to the rest of society, mostly in terms of the wages or salary one earned by being a productive worker. And when the money is transferred to the seller, the purchasing power it represents is also transferred to the seller at the same rate for everyone, namely one dollar per one dollar.

In addition to the problem that the wrong golfers may be getting the rights to play, there are other extensions and implications of using waiting lines to ration demand. It would not take very long for both the buyers and sellers to recognize the wastefulness of having people wait hours to secure the tee time. They will suggest and set up other waiting systems that, at first glance, seem to ameliorate the problem but upon deeper inspection simply serve to move the waiting and competition to another venue. For example, instead of showing up in person to wait one's turn, most golf courses have call-in reservation systems which deserve several comments.[4]

First, many of the advance reservation systems also add a reservation charge which serves as an indirect method of raising the price. As such, it is best interpreted as a halfway move from the subsidized price that requires waiting toward an equilibrium price that does not entail waiting. With the implicit higher price the excess demand is less and the onerousness of the waiting in line system is reduced, but rationing by waiting is not eliminated. Instead of having to beat 182 people per day by lining up early, now one has to "out-compete" fewer people because some drop out because of the higher price.

Second, the problem is not solved, but simply moved from a competition to line up early in person to a competition to call in early to make a reservation. And neither is this competition more fair than using a higher price. People are differently situated with respect to being able to call in at the appropriate time. If the phone lines open at 6:00 a.m., then some less fortunate golfers will have to drop out of the competition if they are unable, for whatever reason, to call at that time. If the phone lines open one week in advance, golfers who cannot set their schedules that far in advance are discriminated against. Golfers who want to play but have to drop out for either of these reasons make the competition easier for the remainder. Others, using regular phones find themselves out-competed by those who are able to take advantage of computerized dialing and answering systems. Why is it more appropriate that the tee times go to more sophisticated "techies" than to those willing to pay higher prices? It seems obvious that competition to earn money by being productive (money which is then used to pay the higher green fee), is more valuable to the rest of society than competition to learn how to manipulate a phone-in reservation system. Perhaps the main advantage of the call-in system is that

[4]There are many call-in systems for golf courses that do not set the price below the market clearing level. These are for the ease and convenience of the golfer and the following problems do not attach to them. When a call-in system is used in conjunction with equilibrium pricing, there is typically no problem getting through on the phone and typically no problem getting a tee time at or near the desired time. When a call-in system is used to ration the excess demand the typical experience involves repeated busy signals at the moment the call-in window opens and then all the tee times except late afternoon partial rounds are taken by the time the caller can get through.

the effort and waiting of the golfers who do secure tee times and the frustration of the golfers who don't are hidden from view.

Third, there are additional costs imposed by a call-in reservation system, costs imposed on the golfers and costs incurred by the golf course. An example of the former is the added uncertainty that attends to having to make a reservation one or two weeks in advance when illness, weather, or sudden demands at work interfere when the reserved time comes up. An example of the latter is the cost of setting up and manning a phone system that will be inundated with multiple calls the second the phone in window opens. Another example of the latter is the requirement to continue answering phones when all the times are taken. If we are talking about an average of 182 people per day calling in to find the course fully booked, we are not talking about a trivial effort by the golf course pro shop and reservation staff.

Fourth, in recognizing the cost and effort involved with the phone-in system, some courses have outsourced this procedure to specialists who are not necessarily on sight. This causes other problems as the following personal experience will attest to. There is one course, which shall remain nameless, which I enjoy playing once a year. The drive is a little longer than I usually have to drive to get to roughly 50 public access courses within a one-hour drive from my house. I called, made a reservation, and specifically asked about the course conditions which I was told were excellent. If I was going to drive the extra distance once per year, I wanted to be sure that there were no temporary holes or special maintenance projects to interfere with my full enjoyment. Upon reaching the first green, imagine my horror to find that the greens had been aerated within the last couple of days. I mentioned this to the pro shop staff at the end of my round and they apologized and revealed that the phone answering company was located in another state. Then the staff even upset me more by asking why I didn't stop after the first hole or even at the turn and ask for a refund if I was unsatisfied. To do so would not really have been a viable option because the investment of time to get to the course had already used up my available time to play for the day. I was begrudgingly offered a refund, but there were bad feelings all around.

The next iteration in attempting to smoothly handle the queue through a reservation system is to remove the human component altogether by implementing a computerized system. This has the advantage of releasing the pro shop staff from continually answering phones by placing all of the burden on the golfer, who can no longer ascertain course conditions or other special daily features like ladies day, seniors day, or a local tournament by talking to an actual person at the course. Internet-savvy golfers will gain at the expense of those less so. However one looks at it, the costs of waiting or the costs of making a reservation have to rise to the level that discourages the roughly 182 golfers per day that are willing to pay the low green fee but cannot get a reservation.

One final problem with queuing systems is that they are potentially susceptible to manipulation by insiders for personal gain. Although few would admit it formally, there are always those who will grant special favors to friends or for a price by allowing them to "cut in line" or otherwise "jump the queue." The pro shop or reservations staff could give priority in granting tee reservations to friends or "regulars" at the expense of the nameless and unidentifiable everyone else. This is worth something to those who stand to gain and they could reciprocate in any number of formal or informal ways. The golfer in question only has to pay $25.00 to the golf course but is willing to pay much more for the reservation. In steps the middleman to skim part of the surplus or rent. It may seem like an innocent favor between friends for which the recipient golfer is grateful and willing to reciprocate monetarily or in kind. Alternatively, it may seem like a justified perquisite of the staff to be able to help out friends. However, it reduces the available number of scarce tee time slots for which everyone else is trying to compete, thus raising the waiting cost of those unconnected to the reservations staff.

Actually, the insider manipulation of the reservation system works similarly to any waiting line system on a number of levels. Some system must be used to distribute the limited number of tee times, and whatever system is used there will be competition to succeed. If a market clearing price is used, then people simply compete productively in the economy to earn money to be able to afford the price. If the price is set too low, then some other system must be established to distinguish those getting the reservations from the rest. If a simple queue is used, the competition is to line up fast, but there is still competition. If a phone or computer system is used, the competition is to become computer savvy to give oneself a speed advantage in getting through to the system. If a personal connection to the staff has to be nurtured, there will be competition to do that. Any of these systems will favor some and disadvantage others based upon individual differences in a person's connections or abilities. But only one of these systems is beneficial to the economy as a whole. Only the use of a market clearing price channels the competition toward productive ends. Do we want to encourage people to productively earn money, or do we want to encourage them to learn how to line up fast, manipulate the phone, or shmooze the pro shop staff?

3.4. Case Study: Securing a Tee Time at Bethpage Black[5]

Lest one think that the above arguments are irrelevant or trivial consider the case of the Black course at Bethpage State Park on Long Island about 30 miles east of New York City. There are a few different paths to obtaining a tee time at this golf course, none of which are as straightforward as walking up and paying a market clearing price would be. No one really can say what the market clearing price would be, but given the documented waiting procedures it is surely higher than the 2012 prices of $65.00/$75.00 (weekday/weekend) for New York state residents[6] who have previously signed up as one of the over 70,000 registered users. Non residents pay twice as much.

Once one signs up as a registered user one is able to use the phone-in reservation system that allows reservations to be made seven days in advance. The call-in lines open at 7:00 pm, but good luck getting through. Consider the simple arithmetic of such a system. Using the numbers from above there could be 456 callers vying for 274 slots. Your chance of being the first caller is one in 456 which is less than one quarter of one percent.

[5]See Branch, J. (2009) Parking All Night at Bethpage, Hoping to Drive, *New York Times*, June 6, Livingstone, S. (2009) Tee Time at Bethpage Worth the Wait . . . Overnight in Parking Lot, *USA Today*, June 18, and Better, C. (2012) How to get Tee Times on Bethpage Black Golf Course, Site of the 2002 and 2009 US Opens, accessed at http://www.golfvacationinsider.com/cr/bethpage-black-golf-course, on May 21, 2012, for documentation of these vignettes.

[6]There is even a senior rate that is lower.

You will most probably get a busy signal at which point you could dial again, and again, and again. Of course, someone is getting through, but the odds favor those with the ability to set up automatic dialing and redialing systems on their computers. At some point the phone lines die down but late afternoon tee times are all that remain and will not accommodate completing all 18 holes. Incidentally, there is a nominal charge for each golfer in a reservation, but it is not enough to bring the total money price up to the market clearing level.

Although there cannot be any formal estimation of the cost of using such a phone-in system, consider the case of Joe Martilotti as reported by Livingstone [2009]. Joe says, "I live right down the road and I've been trying every night, literally for the last two months—using my wife's cellphone, my cellphone, my parent's phone, my house phone. Last week I actually got through." Supposing that Joe tried for one half hour each night, that amounts to over 30 hours. Even evaluated at minimum wage, the cost is more than most people would want to think about. From Joe's story, which formally amounts to one data point, there is no way to tell whether Joe was more or less lucky than average.

The more one calls, the better one's chances of getting through, and there is actually a company that seriously pursues this as its business plan. As Livingstone [2009] reports; "A private enterprise known as N. Y. Golf Shuttle floods the phone lines each night to acquire tee times for those willing to pay $850 per person or $2,000 for a foursome (limo service from Manhattan included)." That's one expensive limo ride. While one cannot blame the company for stepping in as a middleman to redistribute the scarce tee times to those who place extremely high values on them, wouldn't the state's residents actually prefer that this revenue stream accrue to the state park system or be available to generally lower taxes?[7]

For out of state or non-registered users the call-in window starts two days in advance of the desired date of play. Good luck with that, enough said.

There is another onerous waiting system for those unable or unwilling to manipulate the phones or pay someone else to do so. There is a "walk-

[7]Personal conversations with avid golfers in the tri-state area reveal that other, less formal, middlemen are willing to manipulate the phone system to obtain a reservation for between $100 and $200.

up" system for those without reservations. The first six tee times each day and possibly an additional tee time each hour are kept out of the phone system to be allocated on a first-come-first-served waiting system. Actually, the system is a "drive up" system and a car is required (some people having to go so far as to rent a car to use the system). There is a special parking lot with numbered spaces that create the priority. Only the first six numbered slots will actually guarantee play for the next day. In the event that a couple of the cars in slots one to six represent twosomes, then the car in slot seven will be guaranteed one of the first hour's tee times. Typically, the slots fill up the afternoon before for play on the following day, necessitating waiting overnight in the parking lot, and there are written rules concerning requirements about remaining with the car for parts of every hour. Often, people arriving too late on the evening before will actually wait two nights to secure one of the first hour's tee times.

Evidently, the numbered spaces and an honor system are not enough to stop interlopers, queue crashers, or others attempting to steal priority by misrepresentation. Well before dawn, golf employees orchestrate the moving of cars from the numbered slots to the regular parking lot and from higher numbered slots to lower numbered slots. Also, plastic wrist bands are attached with the priority numbers to assure that the integrity of the queue is kept without argument or confrontation. Most people do not want to think about all this unproductive effort on the part of golf course employees and hassle on the part of the consumers. And it is all just to play a round of golf for below the equilibrium price!

The people waiting overnight in this system sometimes consider it part of an experience, a type of badge of honor. An all night poker game with your golf partners, as some have been known to do, is not necessarily a bad experience, although I do not think it would be conducive to playing one's best golf. Regardless, the costs of doing so are significant. One certainly would not want to make it a usual practice to secure tee times in such a fashion. The time and discomfort are costly, and there is an opportunity cost aspect too. As Branch [2009] reported, at least one overnight stayer in space three declined an offer of $1,000 for it. In a similar fashion to the company that floods the phone lines and resells the reservations, when a queuing system is used instead of the price system the incentive still exists for people to circumvent the queue by substituting a money price.

Although these types of private transactions are not documented for obvious reasons (would the seller of such a space report the income to the I. R. S.?), this author is not so naive to think that they never happen.

There is one final wrinkle that is worthy of being mentioned. Supposedly, one additional tee time each hour is kept open to be allocated to the walk-ups. Evidence reported on the golfvacationinsider.com blogs indicates that this policy is open to being manipulated by the golf course staff on a discretionary basis. If the tee times are falling behind schedule, then the open tee time can be skipped over to get the course back on schedule. There are also indications that if a well-connected person or other group of VIP's wanted to play the Black course on short notice, then one of these times could be allocated accordingly. If Michael Jordan wanted to play the Black course, it would be hard to imagine him making repeated calls to the phone reservation system or sleeping overnight in the parking lot. While it is not exactly clear to whom at the golf course payola might be made, one could much more easily imagine Michael Jordan lavishing exorbitant gratuities to the appropriate people as an expression of his gratitude. It is certainly easy to imagine the golf course professionals at the Bethpage State Park facilities securing access for acquaintances or other golf course professionals as a courtesy or as a quid pro quo, or to higher up officials in the State Park system who might expect and demand such perquisites to be supplied by those who, after all, are their employees.

Long waiting lines, wasted time and effort, rounds being distributed to lower rather than higher valued users, and competition in myriad forms to gain an advantage at manipulating the queue or circumventing it altogether, are all on display. And the amounts are not trivial–thirty hours of calling and a two-month delay; 16 to 40 hours of waiting in a parking lot, possibly along with the cost of a car rental; a $1,000 foregone opportunity cost for priority space number three; $850 to an entrepreneurial company acting as a middleman. And consider the bitterly sweet but ultimately frustrating experience of luckily getting through to make a one-week advance reservation only to unluckily find out that when the day comes play is cancelled due to thunderstorms. All this, is just to allow a small subset of citizens, golfers, to save money at the expense of the taxpayers at large.

Chapter 4

Golf Course Waiting: . . . and the Ugly[1]

The previous chapter described how subsidized price policies at publicly-owned golf courses lead to rationing by queuing and how golfers who can negotiate the queuing system with low costs can capture some of the economic rents that are created. This is good for those golfers, however, it is bad for other golfers who cannot acquire tee times or who can only acquire them after incurring a great amount of wasteful, non-monetary queuing costs. It is also bad for the community at large which gets no gains from the wasted effort involved in the rationing by waiting model. This was the good and the bad. The ugly is described in this chapter by documenting how it is possible that subsidized golf course pricing can actually lead to fewer rounds of golf being played at fewer, lower quality golf courses!

The results of fewer rounds at fewer golf courses may seem completely counterintuitive. After all, if municipal golf courses can be built and rounds of golf offered for below market clearing prices without the constraint of having to cover costs, then the possibility arises that a municipal golf course can be supplied in an area that could not support a private, for-profit golf course. It seems like more golf could be supplied in this manner. Unfortunately, the data do not support this interpretation of

[1]The theoretical material in this chapter is drawn from Shmanske, S. (1996). Contestability, Queues, and Governmental Entry Deterrence, *Public Choice*, 86, pp. 1-15. The empirical results come from Shmanske, S. (2004b). Market Preemption and Entry Deterrence: Evidence from the Golf Course Industry, *International Journal of the Economics of Business*, 11, pp. 55-68. The latter is a Taylor & Francis Group journal, see www.tandfonline.com

the building of golf courses. Also unfortunately, the fact that some golfers may gain by receiving low cost rounds of golf, generally convinces the casual observer, who concludes, without stopping to think about the unintended consequences of such policies, that such policies are good for golf and for golfers in general. What can actually happen is far more detrimental to golfers and to golf than is generally understood.

Indeed, golfers may only rarely consider the effect of the municipal golf course's pricing policies on course quality or on the development of other potential golf courses. But let us do so here. By placing oneself in the shoes of a for-profit golf course owner or developer, one would quickly realize that entering into a head-to-head competition with a municipal golf course that can afford to charge low prices and even to make losses that are covered by the community's taxpayers, may not be the best business plan. In a nutshell, the existence of a municipal golf course dissuades, deters, or delays others from entering the industry. And, counterintuitive as this result may seem at first, the data actually support this interpretation of the history of golf course development in the greater San Francisco Bay area. The data also support the conjecture that the municipal courses with such policies can afford to skimp on maintenance and on the provision of quality while still maintaining an advantage due to their subsidized prices.

This chapter proceeds by presenting in section 4.1 a simple heuristic model that establishes the possibility that municipal courses unduly, and perhaps unwittingly, deter the entry of other competing golf courses leading to the result of fewer golf courses overall. And, as a corollary of this main result, this section also shows how course quality may also be lower than otherwise could be. Following this, in section 4.2, grounds for an empirical investigation are laid by describing a data set on the number, type, and vintage of golf courses in a regional geographic area. The statistical regression model is introduced in section 4.3 and the results presented in section 4.4. A short section on golf course quality measurement is in section 4.5. Section 4.6 summarizes.

4.1 A Simple Model of Golf Course Competition and Entry Deterrence

Consider the numerical and diagrammatic model from the previous chapter depicted here as Fig. 4.1. Consider the thought experiment in which two equally situated golf courses in head-to-head competition were serving the market. It is reasonable to expect that they would more or less split the market, and as the diagram shows, they could each break even at point e with a price of $35 per round selling 66,667 rounds per year, while the market cleared at point E on the demand curve.[2]

Now suppose that one of the courses was a municipal golf course and was able to lower its price below the break even level. At first this course

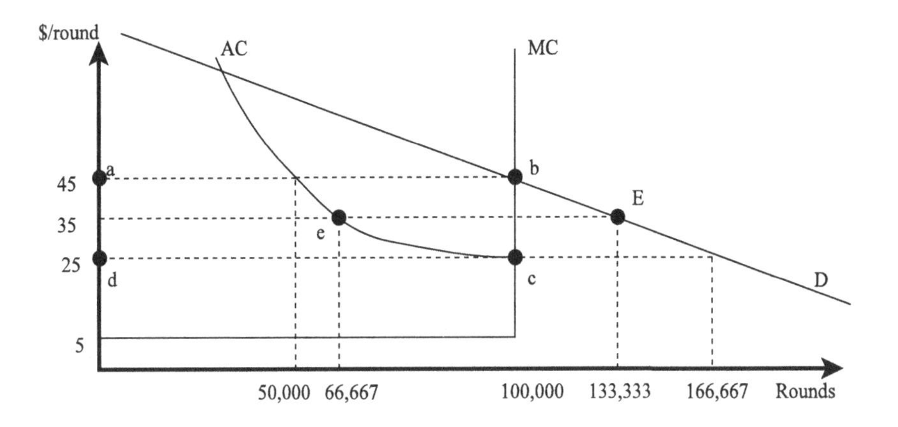

Fig. 4.1 Cost, Demand, Market Clearing Price, and Shortages.

[2]There is a slight rounding error when the whole number demand is used. The exact breakeven point occurs at 66,666 and 2/3 rounds per year. Each of the two courses could be slightly profitable at a price of $40 per round and a quantity of 58,333 rounds, but competition would be expected to lower price to the breakeven level of $35.

would steal customers from its rival, at least to the point where it reached capacity. Instead of splitting the market evenly, the municipal course would sell the first 100,000 rounds because of its lower price which the for-profit course could not match. After selling the first 100,000 rounds, a waiting queue would develop at the municipal course and a waiting cost component would be added to the price as explained in the last chapter. Once, but not until, the waiting cost brought the full cost up to the $35 level of the for-profit course, golfers would also go to the for-profit course. However, there is not enough left over demand to make ends meet after the municipal course captures the first 100,000 demanders. The for-profit course would sell only 33,333 rounds at a price of $35 and would go bankrupt. Its revenues would be $1,166,655 and would fail to cover costs of $2,166,665. The market would end up with only the municipal course, selling 100,000 subsidized rounds with a waiting cost imposed to ration the excess demand as was described in Chapter 3.

It would not do any good for the for-profit course to try to match the lowered price of the municipal course. For example, if both courses charged a price of $30, then market demand would be 150,000 and would be split two ways between the courses. For 75,000 rounds the revenue would be $2,250,000 and costs would be $2,375,000. While the municipal course could withstand the loss with taxpayer subsidies, the for-profit course could not. This is why in a competitive industry the price will not be competed down below the $35 level. The municipal course can lower price below $35 but the result is the closure of the competitor and only 100,000 rounds of golf being supplied. Thus, the result of fewer rounds of golf played at fewer golf courses. As the next sections show, in formal tests the data actually support this interpretation of the development of golf courses.

As a corollary, now consider the incentives facing the operators of the sole municipal course. They are supplying 100,000 rounds for a price below the market clearing level of $45.00. There will be excess demand of an amount depending on the actual price chosen. As the previous chapter shows, the excess demand will have to be rationed in some manner leading to a waiting cost imposed on the golfers. Instead of allowing the line to build up, the golf course operators could decrease the quality of the golf course and still be able to sell out all 100,000 rounds, albeit with a shorter waiting line. By spending less on maintenance or course beautification the experience of the golfer won't be quite as valuable as otherwise and the demand curve will shift slightly to the left. However, since there is excess demand to start with, the course will still be able to sell out all of its rounds. The source of the incentive to skimp on maintenance depends on the actual financial arrangements between the golf course management and the public agency that owns the course. For example, if the city council, county supervisors, or state park directors have tight budgets and other uses for funds, they will push to reduce the budget going to golf course maintenance while arguing that the course will still be able to sell out all its rounds. Alternatively, some arrangements may call for a fixed franchise fee to be paid from the golf course management company to the appropriate public agency. This puts the management company in the position of residual claimant. If so, the management company will still get the same revenues from reaching capacity at the below market clearing prices established by the public agency, but the savings on lower maintenance go directly to its bottom line.

It is fairly typical that municipal golf courses have lower quality than for-profit courses. The golfers know it but are willing to put up with it because of the lower price, while at the same time realizing that the golf course management may be trying their best to maintain a busy course on a shoestring budget. There will also be exceptions to the rule. However, personal experience and anecdotal evidence will have to play second fiddle to the statistical evidence presented in section 4.5 below.

4.2 Data on Golf Course Development in the San Francisco Bay Area

The spatial competition models pioneered by Harold Hotelling [1929] and routinely covered in Industrial Organization textbooks provide the background for the testable implications and the data required to look at golf course development.[3] These intuitively appealing implications include the positive relationship between the population and the equilibrium

[3]See, Hotelling, H. (1929) Stability in Competition, *Economic Journal*, 39, pp. 41-57, and Pepall, L., Richards, D. J., and Norman, G. (2002) *Industrial Organization: Contemporary Theory and Practice*, (Southwestern Publishing, Oklahoma City) chapters 4 and 8.

number of firms, the negative relationship between the population and the equilibrium distance between firms, the negative relationship between the level of fixed costs and the number of firms, and so on. In moving from the computational clarity of the textbook one-dimensional space to the realism of location and competition in a two-dimensional geography, some of the precision is necessarily lost. Nevertheless, an intuitive extension of these models suggests that the same type of relationships should exist in a real world empirical examination. Therefore, in a geographic space with growing population through time, we can test whether golf course entry occurs where and when it should with any degree of statistical regularity.

The golf course industry is perfect for an empirical study of entry and deterrence. First of all, consumption is local. Therefore, the relevant geographic markets are circumscribed and relatively easily measurable. By way of contrast, entry into the wine industry would have a spatial component on the supply side, but the demand would have to consider national, and even international, factors. Second, scale economies in golf are such that there are neither too few nor too many outlets. If only a dozen or so suppliers existed, the data set would not be large enough to carry out a multiple regression analysis. Alternatively, if scale economies are largely absent, then too much entry will take place. Consider, for example the Starbucks chain where the exact location and timing of entry is likely to be dominated by idiosyncratic factors at the micro-geographic level. Studying coffee shop competition requires too much data. Finally, data on the exact location, year of entry, and type of golf course are available for each of the 104 golf courses that were established in the nine San Francisco Bay Area counties from 1893 to 2001. For each of these courses, the relevant market area and the population within it can be measured and correlated to cost, demand, and competitive factors. For example, the relevant, market-area population associated with the entry of any particular golf course is expected to be higher, the greater the number of preexisting courses in the same market.

If there are no incumbents in the relevant geographic market, then we can examine the original entry decision. As the population within a market grows, there comes a point in time when a for-profit firm can enter and expect the present value of future income flows to be greater than the present value of the future costs. Such entry is economically viable.

Preemptive entry can be defined as entry that takes place before this point in time. Ordinarily, a private sector firm will not enter before it is economically viable. A municipal golf course, however, does not have to cover its costs with revenues if it is subsidized by the taxpaying residents. Such entry could be rational for the community if the consumer surplus is counted along with the revenue as part of the benefit stream that has to cover the costs. A statistical look at the growth of the golf course industry should be able to tell us if, in a growing market, the ability of a municipal golf course to incur losses covered by public subsidies allows a municipal course to enter before a profit-seeking, private sector golf course could. But first we need some background on the golf course industry as it applies to our data set.

Traditionally, the golf course industry is divided into three segments: private courses; municipal courses; and daily fee courses. The oldest segment is composed of private golf courses that are owned and paid for by the members, with access restricted to the members and their guests. Typically, a private golf course will only require 200 or so very wealthy families to buy memberships in order to make ends meet. Regardless of what is happening to the total population in an area, in a concentrated pocket of wealthy families a private golf course could be established even if the overall population was low. Therefore, the link between population and entry may be weaker, and the ultimate population required for entry may be lower for private courses than for the remaining two categories.

Municipal courses, allowing play by the general public, are established and subsidized by a variety of local governmental jurisdictions including, but not limited to, cities and counties. These courses are not run to achieve a maximum profit or cash flow. Rather, they are run for the benefit of citizens. Therefore, municipal courses employ a set of preferential prices and policies for city or state residents, and often employ even deeper discounts for senior residents who have it in their interest and find it within their time budget constraints to be heard at city council meetings where pricing policies are hammered out. I have shown in the past that there is a distinct difference in the way municipal courses and for-profit courses use price discrimination.[4] Indeed, resident discounts can be seen as price discrimination in the wrong direction. Profit maximizing, third-degree price discrimination entails charging higher prices to nearby residents who have lower transportation costs and, *ceteris paribus*, higher net demand prices, and who have fewer available substitutes than non-residents who live closer to other courses. All municipal courses have some preferential treatment of residents; none of the for-profit courses do. In some cases the operation of a municipal course might be contracted out to a for-profit management group, in order to be run more efficiently. But even in these cases, the contracts will stipulate the necessary preferential treatment of the city's residents. The important peculiarities of municipal golf course operation for the purposes of this chapter are the facts that they are subsidized, and that they charge lower than profit maximizing prices to a large segment of their customers.

The third category of golf course includes daily fee courses which are privately-owned, for-profit ventures that are open to the public. A subset of this category includes "resort" courses, which are usually located in tourist destinations and connected to a hotel or other lodging. Resort courses give priority to hotel guests, but they will also accept play by the general public on a space-available basis. Being located in tourist destinations, such as the Monterey Peninsula or the Carolina coast, and catering to tourists, it is probable that the link between golf course entry and resident population is weaker for resort courses than for daily fee courses in general. Resort courses, however, are not a factor in this paper's sample of San Francisco Bay Area golf courses. With the possible exception of one or two of the courses in the sample, the daily fee courses rely predominantly on play from nearby residents, and must be located closely to the local sources of demand.

These three golf course types compete with each other for customers, but they are not perfect substitutes. Municipal courses and daily fee courses are close substitutes that are both available to the general public. Private courses might find some competition from daily fee courses, especially the high-end segment of this group, but golfers who join private

[4]See Shmanske, S. (2004a) *Golfonomics*. (World Scientific Publishing Co., Inc. River Edge, NJ) and Shmanske, S. (1998a) Price Discrimination at the Links, *Contemporary Economic Policy*, 16, pp. 368-378.

clubs probably do not play very much golf at municipal golf courses. Another way of saying this is that a wealthy golfer who might consider joining a private club could be satisfied playing golf in posh, and exclusive to the extent of being high-priced, surroundings at the high-end daily fee golf courses, but would probably not be satisfied playing golf at the local municipal golf course. Overall, it is expected that municipal courses and daily fee courses will offer each other the closest competition. The second closest competition will be between private clubs and daily fee courses. The least competition is expected between municipal golf courses and private golf courses.

The golf course industry in the United States as a whole is used as a control variable in the regressions below. Figure 4.2 shows the ratio of population to the number of golf courses in the United States through the

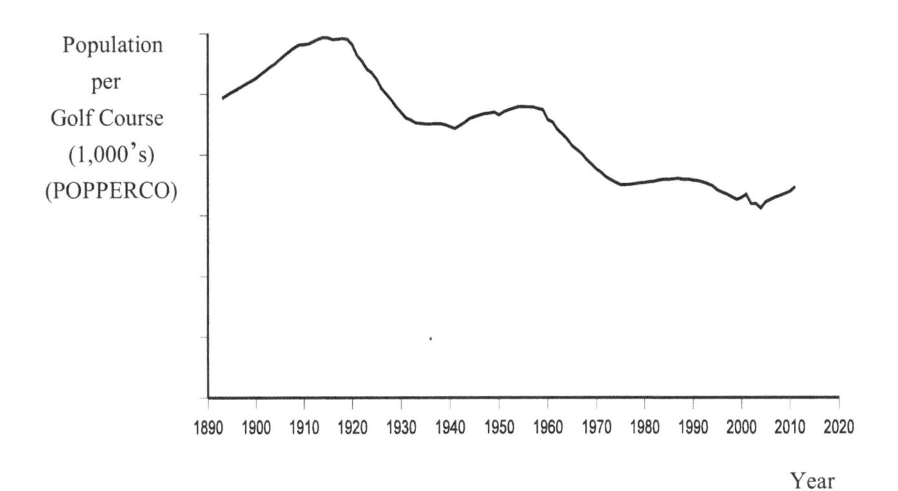

Fig. 4.2 Population per golf course.

years. Through time, the population per golf course has markedly, but not steadily, declined. Turning points in the time series could be associated with a boom period in American golf surrounding Bobby Jones and the 1920's, a setback in golf course construction during the Depression and World War II, and another growth spurt probably associated with pent up wartime demand, Arnold Palmer, and the ushering in of televised golf, and a recent leveling off and retrenchment from an over-exuberant building phase associated with joint housing/golf course development projects.

One further set of stylized facts provides useful descriptive detail about the industry. There is a difference between the character of growth in the golf course industry in the first half of the 20th century and the second half. In this panel of 104 courses in the nine Bay Area counties, 37 were established in the years 1893 to 1949, and an additional 67 courses from 1950 to 2001. Given population growth, and the national trend in the increase in demand (and supply) for golf as evident in Fig. 4.2, this is not surprising. The pattern of growth, however, is also distinct. In the early period, 68% of the golf courses were private, 24% were municipal, and only 8% were daily fee. In the latter period, this rank order is reversed; 39% of entrants were daily fee, 37% were municipal, and only 24% were private. One other difference is very understandable given land use patterns in and around big cities in the 20th century. Of the 37 early entrants, 41% were located in or very near big cities, which for the purposes of this geographical area include San Francisco, Oakland, and San Jose. Of the 67 latter entrants, only 18% were located very close to the three dominant urban areas.

Information was collected for each of 104 golf courses in the nine county San Francisco Bay Area.[5] The data go back to the first course established in Burlingame in 1893, and continue up to and including a course which opened in 2001. For each course, the population within a ten-mile radius of its location was calculated for the year of entry. This is called POP10. Summary statistics for this and all the data vectors appear in Table 4.1. The population within the same area was also calculated for the period five years after entry in order to track the growth of the population. For courses younger than five years old, extrapolations of the actual growth over the available data were conducted to predict the overall

[5]The primary source is Blofsky, E. T. Jr. ed. (2002) *NCGA Golf 2002 Bluebook Edition*, Vol. 22, No. 1. The counties included are: San Francisco, Marin, Sonoma, Napa, Solano, Alameda, Contra Costa, Santa Clara, and San Mateo. Courses of less than 18 holes, or less than a par of 68 were excluded. Other possible exclusions are courses that may have opened, subsequently ceased operations, and been dropped from the Northern California Golf Association (NCGA) data, and courses in the outlying areas of some of the counties that are geographic monopolies in the sense of being farther than ten miles away from their nearest competitor.

Table 4.1 Summary statistics.

Variable	Mean	Standard Deviation	Minimum	Maximum	N
POP10	238,340	238,400	1,803	1,171,000	104
GROWTH	0.23023	0.18877	0.005	1.071	104
P	0.39423	0.49105	0	1	104
M	0.32692	0.47136	0	1	104
DF	0.27885	0.45060	0	1	104
PCOMP	2.1250	1.8988	0	8	104
MCOMP	1.2019	1.3609	0	6	104
DCOMP	0.63462	0.9559	0	4	104
PCAPYDEF	14,537	6,045.7	7,759	33,840	67
POPPERCO	21,886	4,106.9	16,290	31,960	104

Notes: From Shmanske [2004b] Table 1, p. 60. POP10 is the population within a 10-mile radius of a golf course in the year it enters. GROWTH is the 5-year growth in population as a percentage of POP10. POP10 and GROWTH come from the U.S. Census and the California Statistical Abstract, various issues and author's calculations. P, M, and DF are dummy variables capturing the type of course-- private, municipal, or daily fee, respectively. PCOMP, MCOMP, and DCOMP capture the number of incumbent golf courses in the entrant's market area by type--private, municipal and daily fee, respectively. P, M, DF, PCOMP, MCOMP, and DCOMP come from the *NCGA Bluebook 2002 edition*, census maps, and author's calculations. PCAPYDEF is the county per capita income in the year of entry, deflated to 1980-2. PCAPYDEF comes from the California Statistical Abstract, various issues, and author's calculations. POPPERCO is the United States population divided by the number of golf courses in the United States in the year of entry. POPPERCO comes from the Andriot [1983], Graves and Cornish [1998], the National Golf Foundation website, http://www.ngf.org/faq/growthofgolf.html, and author's calculations.

growth for five years.[6] This variable, called GROWTH, is expressed as the five-year cumulative growth in the population as a percentage of POP10.

The use of a ten-mile radius is somewhat *ad hoc*. For the purposes of forecasting demand, the National Golf Course Owners Association recommends counting the population within ten miles of the course and

[6]The calculations are the most tedious part of the research. They are conducted by hand using census maps, a compass, and interpolations from ten-year census counts for the cities and towns falling within the 10-mile radius. Census data get sparser the farther back one goes. However, almost two-thirds of the observations are from 1950 on, where the data are very good, and in the earlier period, many of the courses are near the major urban areas for which the data are also very good.

also the population living between ten and twenty miles from the course and giving twice the weight to the former figure.[7] But their criterion seems to be *ad hoc* as well. The National Golf Foundation has calculated that the average one-way driving distance to the course most frequently played is 10.4 miles.[8] I have successfully used the ten-mile criterion in previous work on the demand for golf, but there are other methods.[9] For example, I used a method of assigning population to courses by determining which course is the closest from every location in the Bay Area.[10] This method closely captures the spirit of the formal spatial competition models where consumers shop only at the nearest outlet. The ten-mile radius rule, by contrast, counts some of the population more than once if golf courses are located within ten miles of each other, thereby assuming overlapping market areas. In my view, overlapping areas make sense because most golfers do exhibit a taste for diversity by not exclusively playing the nearest course. If the choice of radius is made too large, (for example, large enough to cover the whole geographic area in question), then all courses entering in the same year would be indistinguishable from each other vis-a-vis the population measure. This is tantamount to saying that the relevant market for each course is the whole Bay Area which we know is not true.

The type of golf course, private, municipal, or daily fee, was recorded for each course as was the number and type of other golf courses falling within the same ten-mile radius and predating the establishment of the course in question. Thus, for each course, we know its type, and the number of incumbent golf courses of each type that comprised the local competition for the entering course. For example, MCOMP is the number of municipal courses that already exist in a ten-mile radius surrounding the new entrant. DCOMP and PCOMP are analogously defined for daily fee courses and private courses. A dummy vector, DF was created with the value of one if the new entrant course was a daily fee course, and a zero

[7]*Marketing Plan Manual*, National Golf Course Owners Association, 1987.

[8]*Golf Consumer Profile 1989 Edition*, National Golf Foundation, August 1989.

[9]See, Shmanske, S. (2004a) *Golfonomics*. (World Scientific Publishing Co., Inc., River Edge, NJ), and Shmanske, S. (1998a) Price Discrimination at the Links, *Contemporary Economic Policy*, 16(3), pp. 368-378.

[10]See, Shmanske, S. (1999) The Economics of Golf Course Condition and Beauty, *Atlantic Economic Journal*, 27(3), pp. 301-313.

otherwise. Similar dummies were created for municipal courses, M, and private courses, P. Then, these dummies were interacted with the variables containing the number of incumbent competitors of each type. For example DFXMCOMP has the number of preexisting municipal courses within a ten-mile radius of the entering course if the entering course is a daily fee course, and zero otherwise. The extent to which municipal courses deter the entry of daily fee courses will be captured by the coefficient of this variable. Because there are three types of entrants and three types of incumbents, there are nine such variables altogether.

Population figures for the United States as a whole and for the number of golf courses that exist in any one year are also collected. These data series form the basis for Fig. 4.2. National trends in the popularity of golf provide a backdrop for the development of golf courses in the San Francisco area. This factor can be controlled for by including the population per golf course in the country as a whole (called POPPERCO) in the regression equation. If golf is more popular nationally, this may influence the timing of the development of golf courses in a local area.

Per capita income by county for the year and county of each entrant was also collected, and then deflated to 1980-2 dollars in the vector, PCAPYDEF. Useable data went back to 1950 so the model was run separately on the 67 instances of golf course entry since 1950. PCAPYDEF was decidedly insignificant, and the other coefficients were unchanged, so the variable was dropped and the model was run on the full sample of 104 golf courses in the reported results. It seems reasonable to think that suitably measured income should show up as a significant determinant of the demand for golf, and, therefore, the golf course entry decision. Evidently, however, county-wide per capita income is too rough a measure to capture this effect.

One final dummy variable, BIGCITY, was constructed. This variable has a value of one if the POP10 variable includes the population of San Francisco, Oakland, or San Jose. This variable works as a proxy in two ways, and as a measurement correction, to be explained below.

4.3 A Statistical Regression Model of Golf Course Entry

The model presumes that a golf course enters when the discounted value of the future stream of expected benefits is greater than the discounted value of the future stream of expected costs. Among other things, the benefits depend on the demand to golf at the locale in question which depends, in turn, on the relevant population. Therefore, as population grows in an area, the establishment of a golf course becomes more likely. At some point in an area's growth, the scales tip and the golf course enters. The population at this point is captured by POP10. Therefore, using POP10 as the dependent variable in a multiple regression equation, we can essentially find out what factors cause the scales to tip in favor of entry sooner, and what factors cause the scales to tip later. Therefore, the following regression equation was estimated.

$$\begin{aligned} POP10 = {} & b_0 + b_1 M + b_2 DF + b_3 DFXMCOMP + \\ & b_4 DFXDCOMP + b_5 DFXPCOMP + b_6 MXMCOMP + \\ & b_7 MXDCOMP + b_8 MXPCOMP + b_9 PXMCOMP + \\ & b_{10} PXDCOMP + b_{11} PXPCOMP + b_{12} GROWTH + \\ & b_{13} BIGCITY + b_{14} POPPERCO + e \end{aligned} \tag{4.1}$$

Each right hand side variable and its expected effect will be explained in turn. The coefficients of the dummy variables, M and DF, capture the effect of the type of course on the population required for entry. The dummy for private courses is omitted so the effect of private courses is captured in the constant term along with the average of all other omitted effects. The coefficients of M and DF capture any extra population (over that required for a private course to enter) necessary for these types of course to enter. It is expected that private courses can enter with the lowest necessary population because private courses do not require a broad-based support from the population at large. Instead, private courses require a large commitment from relatively few families. Therefore, significant positive coefficients are expected for M and DF, indicating that more population is needed to support the entry of these types of courses. Alternatively, if municipal courses, owing to subsidies, preempt the entry of private and daily fee courses, then the coefficient of M should be negative and significant. If municipal courses preempt only daily fee courses, then the coefficient of M could be positive but should be significantly less than the coefficient of DF. Other preemption hierarchies could also be determined based on the signs and magnitudes of the coefficients of DF and M.

The next nine variables capture the entry deterring effects on the differing types of entering golf course by the differing types of incumbent golf course. For example, the coefficient of DFXMCOMP captures the extra population that a daily fee course requires to enter the market for each municipal course that is already established nearby. Alternatively, the coefficient of DFXDCOMP captures the extra population that a daily fee course requires to enter the market for each incumbent daily fee course. There are three types of entrants and three types of incumbents, making for nine coefficients in all.

To the extent that all golf courses regardless of type are substitutes then all nine coefficients should be positive and significant. In the limiting case of perfect substitution among the three types, all nine coefficients should be equal. In the other limiting case that has the three types of golf course providing distinct, unrelated goods, only three of the nine coefficients–the ones capturing the effects of existing competition on entry of golf courses of the same type–should be significant. But neither of these patterns is expected given the nature of the three types of golf course.

I expect the least competition to be felt between the municipal courses and the private courses. Therefore, the coefficients of MXPCOMP and PXMCOMP are expected to be the lowest of all nine coefficients or to be statistically insignificant. The strongest competition should exist between municipal courses and daily fee courses. Thus, the coefficients of DFXMCOMP, DFXDCOMP, MXMCOMP, and MXDCOMP should be positive and significant, measuring the extra population required to enter a market for each incumbent of these two types. Private courses will have a deterrent effect on other private courses, so PXPCOMP should be positive and significant, but perhaps lower than the preceding four coefficients if private courses enter based on concentrations of wealth and not on the overall population level. Finally, to the extent that there is effective competition or substitution between private courses and daily fee courses, the coefficients of PXDCOMP and DFXPCOMP will be positive and significant.

Testing the entry deterrence proposition requires a closer look at the estimated coefficients of DFXDCOMP and DFXMCOMP in particular. The conjecture is that municipal courses, by keeping prices low through taxpayer subsidies, effectively make it harder for new, for-profit competition to enter. It is obvious that an existing daily fee course will make it harder for a new daily fee course to enter; this effect is captured by the coefficient of DFXDCOMP. If it is even harder than this to enter against an incumbent municipal course, then the coefficient of DFXMCOMP should be greater than the coefficient of DFXDCOMP. This comparison of coefficients is the main test of the entry deterrence hypothesis.

Before moving to the results there are three other control variables to briefly explain. The first, GROWTH, is interpretable as a rational expectations effect. The population that will form the customer base for a golf course need not be in place at the outset, if it is expected to move in shortly. The faster the expected growth in the population, the lower the population has to be now in order for a course to enter. The actual cumulative growth in population over the five years subsequent to entry is used as the proxy for the expected growth of the population. The coefficient is expected to be negative.

The dummy variable, BIGCITY, is included as a possible proxy for two separate effects that would seem to be positively correlated to urban population concentrations, and for a third effect that has to do with population measurement. First, BIGCITY acts as a proxy for land values. Land values are higher in and near large cities. Land values are relevant to the golf course entry decision because acquisition of the land is a major cost of golf course development. If the land cost is higher, as it is in or near a big city, the amount of the population has to be higher before the demand will be high enough to justify entry. Therefore, the coefficient of BIGCITY should be positive.

The second reason for including BIGCITY, and acting in the same direction as the first, is the fact that it is in urban environments where pockets of poverty develop most significantly. Since golf is surely not a necessity, it is likely that the portion of the population below the poverty level does not effectively add to the population that a golf course can count on to supply paying customers. This means that the POP10 variable is effectively too large when it includes urban populations. The BIGCITY dummy will also capture this effect with a positive coefficient.[11]

[11] In another paper, I was able to subtract out the portion of the population below the poverty line (and double the amount of senior citizens who strongly demand golf) when dealing with only one year of population data. The results were better with the corrected population data which was used as an independent variable in that paper. The current study has potentially 104 years of data to correct, and the definition of the poverty line changes, making the same approach impossible. See, Shmanske [1998a].

The third reason to include BIGCITY is that it provides a slight technical correction for a possible undercount of rural population in the early years of the data. Population in the urban and developed areas was easy to capture even going back more than 100 years. In rural areas, however, the farther back in time one goes, the more people live in unincorporated areas of the county, and are, therefore, harder to attribute to any one ten-mile radius area. For this reason it is possible that the POP10 figures are undercounted in rural areas relative to the urban areas. A positive coefficient on BIGCITY will also pick up this effect.

The final right-hand-side variable is POPPERCO, which is a proxy that is negatively correlated to the nationwide popularity of golf. If the California experience is similar to the rest of the nation's, then the coefficient of POPPERCO should be positive. That is, a higher value of POPPERCO signifies a lower demand for golf, and consequently a higher level of POP10 before entry will occur.

4.4 Results on Golf Course Entry

Equation (4.1) was estimated with ordinary least squares and, given the presence of heteroskedasticity, the results are reported with White's [1980] robust variance-covariance matrix. The results are reported in Table 4.2.

Overall, the fit is good as evidenced by the adjusted R^2. Ten of the coefficient estimates are significantly different from zero in the expected direction, and of reasonable magnitudes. There are no instances of significant coefficients with the wrong sign.

The BIGCITY and GROWTH control variables perform as expected. For each percentage point of cumulative population growth over the next five years, the current population base that would attract entry falls by 1,146.5 people. Meanwhile, locating in or near a big city requires a significantly higher nearby population compared to a suburban or rural location. There are three possible interpretations of this result, among which this data cannot separate, namely, (1) that land prices are higher in urban areas, (2) that urban poverty translates into a lower effective demand for golf, and (3) that rural population counts in POP10 are relatively understated compared to urban population.

Neither POPPERCO nor its inverse turn out to be significant. Interest in golf may be one more item on a list of items where the San Francisco BayArea refuses to conform to the general population.

Table 4.2 Estimation of Eq. (4.1).

Dependent variable: POP10 Adjusted R^2: .823 N = 104

Variable	Coefficient	t-statistic
Constant	-135,201	-1.137
M	60,101	1.364
DF	44,237	0.959
DFXMCOMP	76,734	5.880
DFXDCOMP	39,650	2.383
DFXPCOMP	14,567	2.432
MXMCOMP	68,607	3.549
MXDCOMP	69,670	2.535
MXPCOMP	19,838	1.199
PXMCOMP	39,665	2.031
PXDCOMP	142,285	3.909
PXPCOMP	25,434	2.632
GROWTH	-114,650	-2.142
BIGCITY	282,699	9.798
POPPERCO	5.913	1.365

Notes: From Table 2, p. 65, [2004b]. Method--Ordinary least squares with White's robust variance-covariance matrix.

The tests of market preemption involve the comparison of the constant term with the dummy variables for municipal courses, M, and daily fee courses, DF. The coefficients of all three of these variables are statistically insignificant, so we must conclude that there is nothing going on. For the first entrant into any market, it does not appear that any one type of course has an advantage. The only story that can be told about first entry comes from the actual history of the golf course industry in the area. Examination of the data indicates eleven cases in which a private course was the first to enter a specific ten-mile radius area, two such cases for municipal courses, and two such cases for daily fee courses. This pattern is probably due to the fact that the early industry was dominated by private courses when golf was a rich man's game. A private club would ordinarily be the first golf course in any given area. This is probably less true today now that golf has filtered down to middle and lower income levels. If we take out the private golf courses and look only at the public-access courses, the history reveals that municipal courses were the first to enter eight times, while daily fee courses were the first in five cases. Eight to five is not what anyone would call a smoking gun with respect to preemptive entry on the part of municipal courses. While interesting, these stylized facts do not carry over to specific tendencies when other factors are controlled for in the multiple regression setting. The bottom line is that there is no credible evidence that subsidized public golf courses allow for more golf by being established for the benefit of residents earlier than economically viable.

The tests of entry deterrence are another story. The coefficients of all nine variables capturing the effects of preexisting competition on entry are positive, and eight of the nine are statistically significant. The estimates range from 14,566 extra persons required per existing course to 142,285 per course. It is clearly the case that the more preexisting competition there is, the larger the market has to grow before new entry takes place.

Comparison of these estimates by type of course is also illuminating. A priori, we expected there to be the least effect between private courses and municipal courses. Indeed, the one insignificant estimate occurs for the effect of existing private courses on the entry of a municipal course. Interestingly, however, one of the smaller significant effects shows up for the effect of each existing municipal course on the entry of a private course. Each municipal course increases the population required for entry of a private course by 39,665 people. Evidently, golfers who are the target market for a private course are somewhat satisfied to play the local municipal course.

The relationship between private courses and daily fee courses is even more interesting. There is only a small effect of existing private courses on the entry of daily fee courses; an extra 14,567 people are required for each incumbent private course. The effect in the other direction is almost ten times as large. For each incumbent daily fee course, a private course requires an additional population base of over 140,000 people within a ten-mile radius. Can these estimates be reconciled? I think so, given the nature of private and daily fee courses. A private course essentially skims the top portion of the market, getting enough revenue to operate from a relatively small number of people. After this skimming, the bulk of the market remains unserved because of the exclusive nature of private clubs. Thus, the bulk of the market is still available to patronize a daily fee course. In an area with growing population, the initial establishment of a private golf club, delays the entry of a daily fee course for only a short period of time. This effect, however, is not symmetric. The open-access policy of daily fee courses means that if a daily fee course exists, wealthy golfers are not precluded from playing golf. Therefore, these wealthy golfers, who are the prime targets of cream skimming by a private club, are not as desperate to join a private club. Simply put, on the one hand, an existing private course with its exclusive policies, is not a substitute for a new entrant daily fee course to the vast majority of the population who are non-members of the private club. On the other hand, an existing daily fee course with its open-access policies is at least a partial substitute for an entering private course. It stands to reason that the deterrent effect of a daily fee course on a potentially entering private course is greater than the deterrent effect of a private course on a potentially entering daily fee course. To a lesser extent the same pattern was evident in the relationship in the preceding paragraph between private courses and municipal courses, although the effects were not as high and not as statistically significant.

The best tests of competition and entry deterrence come from the comparison of municipal courses and daily fee courses. For a municipal course to enter the market seems to require about 69,000 more people for each existing public access golf course regardless of type. Evidently, the pressure on city councils to develop a municipal course is lessened by the existence of another course in the area at which the city's residents could play. And it does not matter whether that course is a daily fee course or a municipal course. This makes perfect sense. From the point of view of city A, neighboring city B's municipal course will look just like a nearby daily fee course, in that neither type grants discounts or offers preferential treatment to the residents of city A. The coefficients of MXMCOMP and MXDCOMP should be equal, and they are, in both the sense that the difference between them is not statistically signifcant, and also in the sense that the coefficient estimates themselves (68,607 and 69,670) are extraordinarily close.

Finally, however, consider the effect that existing public access courses have on the entry of daily fee courses. Here is where we obtain a test of the unwitting entry deterrence hypothesis. If it is harder to enter a market against an incumbent municipal course than an incumbent daily fee course, then the coefficient of DFXMCOMP should be greater than the coefficient of DFXDCOMP, and it is. The difference in the coefficients is slightly over 37,000, and the difference is statistically significant.[12] The explanation for the difference is clear. It will be harder to compete and make ends meet against a taxpayer subsidized seller of golf at discounted prices than against another profit-seeking golf course that also has to make ends meet. The municipal golf course is clearly not trying to forestall or blockade entry, but its policies have exactly that effect. The evidence is striking, municipal courses do not enter earlier to provide golf where there otherwise would not be any, but once they are built, the entry of daily fee golf courses is slowed considerably. This establishes the results of fewer rounds of golf on fewer golf courses. We now move to the evidence on course quality.

4.5 Results on Golf Course Quality

There is no obvious objective measure of golf course quality. Nevertheless, golfers can form opinions about the quality of their golfing experience. Therefore, a subjective measure of golf course quality can be obtained by asking the golfers. This is precisely what I did in a sample of 46 municipal and daily fee golf courses in the San Francisco Bay Area in 1993. I interviewed a volunteer from each of 20 groups of golfers as they finished the last hole on each course, leading to over 900 separate interviews. Among other questions, I asked the golfers to rate the overall condition of the golf course on a one-to-five scale with one being poor and five being excellent. The average of these ratings for municipal courses was 3.47 which was lower than the average of 3.61 for daily fee courses. The distributions of the average ratings by course type largely overlap, but at least on average, as far as the point estimate goes, municipal courses are rated as being in worse condition than daily fee courses.[13]

[12]The t-statistic for the extra effect of a municipal incumbent on a daily fee entrant, over and above the combined effect of municipal and daily fee incumbents is 2.189.

[13]For more detail about the interviews see Shmanske [2004a] especially Chapter 4.

4.6 Summary and Discussion

The results of this research are not favorable with respect to municipal golf. Municipal golf courses might help golfers in areas that cannot otherwise support a golf course, by entering even when revenues do not cover costs. In this data, however, there is no evidence that municipal golf courses enter before or in areas of less demand than daily fee courses do. There is evidence, however, of a deterrence effect whereby fewer daily fee courses enter a market once a municipal course is established. Overall this means that there are fewer golf courses and less availability of golf than there otherwise would be.

On another level, the quality of the golfer's experience may suffer when the crowded, underpriced municipal course can lower its quality and still reach capacity owing to the dearth of other nearby competing courses due to the deterrence effect. In a survey of San Francisco Bay Area golfers the average condition of municipal golf courses was rated below that of daily fee courses.

Privatization is the cure for these pathologies. In the long run there will be more golf courses, better golf courses, and more golf. In the short and long runs there will be a lower tax burden. These are all gains. There are, however, clearly identifiable losers in the short run from privatization, namely those who benefit from the discounted prices, especially those who have learned how to game the queuing or reservation system to get the benefits of the low prices without the costs of the rationing-by-waiting. At the typical municipal golf course, these are the senior citizens and the other "regulars" who play every week, sometimes multiple times per week, and who will show up at city council meetings to defend the status quo and their own preferential treatment.

One final comment is appropriate. Although this chapter focuses on golf courses, the theoretical argument is not specific to golf. Publicly subsidized providers can deter the entry of for-profit providers in other markets as well. Public schools, public parks, public tennis courts, public libraries, public mail service, public transportation, public ambulance service, public hospitals, public emergency rooms, public disaster relief, and public dispute resolution mechanisms, all have taxpayer subsidized prices, and occasional or chronic problems of excess demand that force the use of non-price rationing schemes, and may deter welfare-enhancing entry by private producers. The golf course industry has provided an empirical testing ground for the basic theoretical argument. Based on the results for golf, privatization of these other industries may be equally beneficial and merits consideration.

Chapter 5

Consistency or Heroics[1]

PGA TOUR professional golfers play in tournaments where the payment structure to the winners is decidedly nonlinear. It is heavily weighted to reward the top performers more handily, for example, the winner typically receives 18% of the purse, second place receives 10.8% and so on down to 0.2% for 70th place. This payment structure disproportionately rewards superlative performance and brings forth more effort as predicted in the tournament compensation model of Lazear and Rosen [1981] and statistically verified in a couple of papers by Ehrenberg and Bognanno [1990a, b]. It also brings about the possibility that some of the hallmarks of the game of golf, namely, consistency and steady play, will be eclipsed by one-time heroics or flash-in-the-pan, "hot hand" performances.[2] Indeed, the payment structure brings up the possibility that the PGA TOUR is disproportionately rewarding one-time, exceptional performances rather than consistent steady play.

Although it is not immediately clear what PGA TOUR officials can do to affect consistency or heroics, the issue is important to the marketing of professional golf. On the one hand, for example, if mean performance were the only thing that mattered (that is, if the variance of individual golfer's

[1]The material in this chapter is drawn from Shmanske, S. (2007) Consistency or Heroics: Skewness, Performance and Earnings on the PGA TOUR, *The Atlantic Economic Journal*, 35(4), pp. 463-471.

[2]The object in golf is to achieve the lowest score possible, so that having a hot hand is also referred to as "going low." Hot hand hypotheses have been explored in other sports, notably basketball, but are measured somewhat differently. See Camerer [1989] and Stone [2012].

scores approached zero), then the outcome of a tournament would essentially be a foregone conclusion. While some fans would still pay to see what amounted to a display or exhibition of exceptional skill, there would be no interest in the sport through the avenue of uncertainty of outcome. There would be no value created for fans wishing to root for their favorites and gambling markets would largely disappear. On the other hand, if a large variance in golfer's individual performances (or a large negative skewness) was of predominant importance, then different golfers would win each week and a different set of golfers would be among the leaders in each tournament. Such "competitive balance" might be desirable for team sports organized in leagues, but the issue is an unsettled one [Schmidt and Berri, 2001; Szymanski, 2001; Zimbalist, 2002; Humphreys, 2002; Sanderson and Siegfried, 2003; Fort, 2003; Kahane, 2003]. In golf, however, it might be detrimental. Name recognition would suffer and the lack of one or a few superstars might cause fan interest to wane [Rosen, 1981]. Indeed, it has often been documented how television ratings suffer when Tiger Woods is not in contention [Fitzgerald, 2006]. Is it consistency in the sense of low mean scores, low variances, and insignificant skewness that matters in professional golf? Or is it heroics in the sense of a high variance and a large negative skewness in the distribution of a golfer's scores that golfers cash in on? Or is it both?

These questions have practical implications outside of sports economics. Consider our college which gives monetary awards for superior teaching and superior research that faculty members can win once every five years. There is little doubt that those in possible contention for the awards give extra effort during the years in which they are eligible. This, of course, is the whole point of giving such recognition to outstanding faculty members. However, there is the possibility that previous winners of the awards could slack off in the effort department and rest on their laurels during years in which they are not eligible. A dose of superior effort and performance now and then might not be preferable to consistent, steady effort over the long term. Similarly, employees may put forth intermittent effort to compete for employee of the month or salesman of the year honors and awards and slack off at other times if compensation is too weighted towards only top performers.

In one sense we know the answer to these questions. Tiger Woods became a household name because of his consistently superior play, and there have always been superstars from Bobby Jones and Walter Hagen through Ben Hogan to Arnold Palmer, Jack Nicklaus, and Tom Watson, to recent stars, Woods, Phil Mickelson, Ernie Els, Sergio Garcia, and Rory McIlroy. However, the extent to which golfers achieve stardom (and earnings) through steady play and low mean scores, versus through mediocre play with a large variance that allows one to contend for first prize roughly half of the time, versus through relatively poor play on average with a large negative skewness that leads to a small number of exceptional performances with large paydays, is still an empirical question. Ultimately, there need not be one path to financial success. Which golfers achieve success through which of the above paths can be identified by an examination of the data.

I use the performances of the top 100 money winners on the official PGA TOUR money list for the year 2002 as the basis for the research. The next section provides a review and visualization of the mathematical concepts of variance and skewness. Following that, the raw data, summary statistics, and some interesting lists of leaders in each category are described. Section 5.2 also describes an important transformation of each golfer's scores to account for the difficulty of each tournament's golf course and the strength of the field of competitors. The third section reports the results of regression analyses that confirm the importance of the measurement protocol adopted with respect to each golfer's scores, and show the independent effects of mean, variance, and skewness on the golfer's earnings. The final section summarizes, places the results in the context of other research on golfer's earnings, and sets the stage for extensions.

5.1 Variance and Skewness

Table 5.1 lists three possible distributions of golfer's scores relative to par. Each distribution has 20 scores. The average, or mean, score in each distribution is zero, that is, par. But the distributions differ in their variance and in their skewness. To review, the variance is a measure of the spread of the distribution of data around the mean and is essentially the average of the squared differences of each observation from the mean. The differences from the mean can be negative or positive, but since each

Table 5.1 Distributions to illustrate variance and skewness.

Observation	Distributions A	B	C
1	-3	-4	-5
2	-2	-3	-3
3	-2	-2	-2
4	-1	-2	-2
5	-1	-2	-2
6	-1	-1	-1
7	-1	-1	-1
8	0	-1	-1
9	0	0	0
10	0	0	0
11	0	0	0
12	0	0	1
13	0	1	1
14	1	1	1
15	1	1	1
16	1	2	2
17	1	2	2
18	2	2	2
19	2	3	3
20	3	4	4
Mean	0	0	0
Variance	2.21	4.21	4.74
Skewness	0	0	-0.34

difference is squared in the formula, the variance will always be positive. In mathematical parlance the average of the squared differences from the mean is called the second moment. Skewness essentially extends this idea to the third moment around the mean, that is, the average of the cubed differences of each observation from the mean.[3] As above, differences from the mean can be either positive or negative, and since cubing these differences will preserve the positive or negative sign, skewness can be positive or negative.

Variance captures the extent to which the individual observations spread out far from the average of the distribution. In colloquial parlance, the longer or fatter are the "tails" of the distribution, the greater the variance. Meanwhile, skewness captures the asymmetry of the distribution, that is, the extent to which the tails of the distribution are longer or fatter on one side than the other. A negative skewness means the tails are longer or fatter on the left, that is, for observations below the mean. In the golf example, a significant negative skewness indicates the ability to have a number of scores that are under par by a large amount, at least more so than the number of scores that are over par by a large amount.

Perhaps this is an area where a picture is truly worth a thousand words. Consider Fig. 5.1 which shows histograms for the three distributions of scores listed in Table 5.1. Distribution A is pictured in the top panel. It is a symmetric distribution, centered on zero, with a variance of 2.21 and a skewness of zero. Now compare panels A and B. The distribution in panel B has longer, fatter tails as captured in its greater variance of 4.21. It is still symmetric with a zero mean and a zero skewness. Finally, compare panels B and C. The distribution pictured in panel C is asymmetric, with a longer negative tail than distribution B and an identical positive tail as in B. Distribution C is said to be skewed to the left or said to possess a negative skew and, indeed, the numerical value of the skewness is -0.34.

Many economic and statistical analyses consider the averages and variances of a variety of distributions. By comparison, the use of skewness in economic and statistical analysis is rare. Perhaps there are not enough data to capture enough different distributions so that comparisons of their

[3]The precise formula for skewness need not be reproduced here. It is based on the third mathematical moment but also includes a scaling factor based on the second mathematical moment.

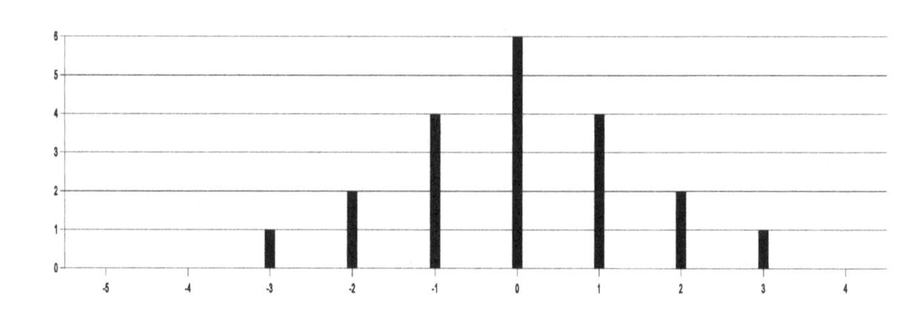

Panel A

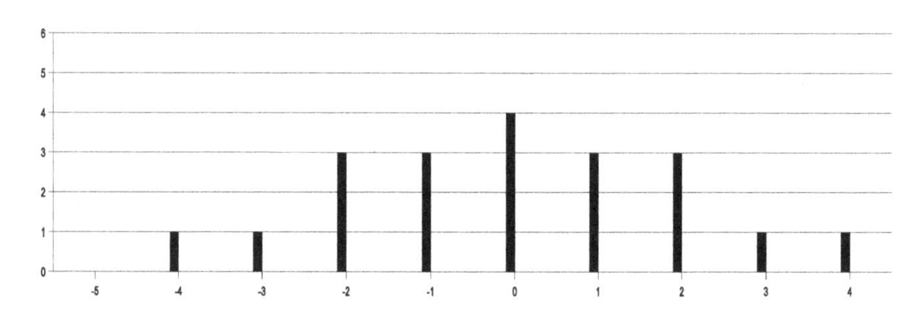

Panel B

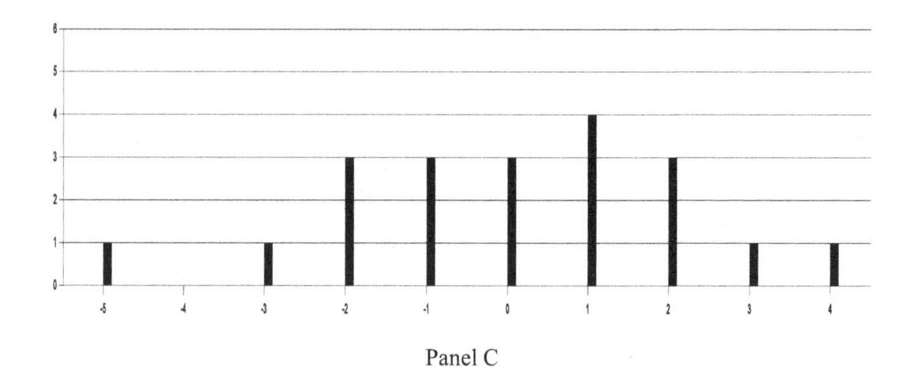

Panel C

Fig. 5.1 Histograms of distributions A, B, and C.

skewness would be informative. Here is another example of the abundant data in sports economics coming to the rescue. By comparing the scoring distributions of 100 different golfers and calculating the skewness of each, we can determine the extent to which skewness varies, and the extent to which earnings outcomes respond to the differently skewed distributions of scores.

5.2 The Data

Each golfer's final score and standing in each of the 49 2002 official PGA TOUR tournaments is listed in the *2003 PGA TOUR Media Guide.* I use the results of the top 100 money winners in 47 of these tournaments. Two tournaments are omitted because they do not use the standard scoring system where each and every stroke counts.[4] The summary statistics are listed in Table 5.2. For each of the top 100 golfers the raw data consists of

Table 5.2 Summary statistics.

Variable	Mean	Standard Deviation	Minimum	Maximum
SCOREAVE	70.704	0.549	68.56	71.85
NUMBER	26.890	4.151	11	34
MONEY	$ 1.52M	$ 0.932M	$ 0.676M	$ 6.913M
EARNPER	$ 59,834	$ 46,721	$ 22,860	$ 384,000
VARIANCE	5.219	1.596	1.354	9.199
SKEWNESS	0.406	0.503	-0.553	1.950
RELMEAN	0	0.717	-2.826	1.282
RELVAR	2.751	1.043	1.309	6.580
RELSKEW	0.363	0.579	-0.630	2.174

[4]One of these tournaments is a match play, single elimination tournament; the other uses a modified Stableford scoring system in which low scores relative to par on individual holes are more valuable than in regular stroke play.

the golfer's average score per 18 holes for each tournament that the golfer entered, and for the year as a whole [SCOREAVE], the number of tournaments entered [NUMBER], and the golfer's total money earned during the year measured in millions of dollars [MONEY]. The dependent variables used will be the earnings per tournament [EARNPER = MONEY/NUMBER] and the logarithm of earnings per tournament [LNEARNPER].

Professional golfers make choices about which and how many tournaments to enter based on the golf course characteristics and location, on which other golfers are likely to compete, on the size of the prize fund for the particular tournament, and on injury, fatigue, and other idiosyncratic personal factors.[5] Because of the fact that the top 100 golfers are not always competing only against each other on the same course, adjustments to the raw scores are required. It is straightforward to calculate each golfer's average score, and the variance [VARIANCE] and skewness [SKEWNESS] around that average, directly from the golfer's actual scores, but the score on any one golf course is not exactly comparable to the score on another golf course. The average score across all golfers in one tournament will differ from that in another due to course length, course difficulty, weather variability, *and* the quality of the golfers entered in the tournament.

To adjust for these influences I estimate what is tantamount to a two-dimensional fixed effects model that simultaneously controls for each individual and each individual tournament venue. Essentially, a regression is run with the raw scores as the dependent variable, and dummy variables for each of the 100 golfers and each of the 47 tournaments. The estimated coefficients of the golfer variables give the relative performance of each golfer controlling for the difficulty of the tournaments they entered. Meanwhile, the estimated coefficients of the 47 tournaments give the difficulty of each tournament controlling for the strengths of the golfers that entered the tournament. The 47 individual tournament course dummy variables become the adjustment factors for each tournament. These adjustments range from 67.51 for the Greater Milwaukee Open to 74.08 for the U. S. Open. The coefficient is subtracted from the golfer's actual score

[5]A closer look at the entry decision is covered in Chapter 7.

to compute the golfer's relative score (on an 18-hole basis) for the tournament.[6] So, for example, someone shooting a 67 in the Greater Milwaukee Open is only 0.51 strokes lower, that is better, than average. Meanwhile, someone shooting a 67 in the U. S. Open gets a negative 7.08 as a relative score, that is, more than 7 strokes lower or better than the average. These adjustments are made to each of the 100 golfer's scores for each of the 47 tournaments resulting in 100 different distributions (one for each golfer) of relative scores.

These relative scores are comparable across tournaments, therefore, the distribution of these scores can be studied. The mean, variance, and skewness of the distributions are calculated. These characteristics of each golfer's distribution of relative scores are called RELMEAN, RELVAR, and RELSKEW, and are the independent variables of interest in the main regressions below.

The extra work entailed in this data transformation should be worthwhile. Consider Tiger Woods as an example. Woods was the leading money winner in 2002 (overall and per tournament) with official earnings of $6,912,625, earned in only 18 tournaments.[7] Woods also had the lowest SCOREAVE of 68.56 strokes per round, over three strokes lower than the *highest* average of those in the top 100, Heath Slocum at 71.85 strokes per round. But this measure does not do Woods justice because he typically plays in the hardest tournaments with the highest scores and eschews tournaments on easier golf courses where everyone shoots really low scores. Justice is done by using the relative score average as described

[6]This is superior to simply using the average score in the tournament because the strength of the field differs for each tournament. For example, if Tiger Woods does not play, the average score for a tournament is likely to be a little higher than if Woods did play. This means that the adjusted relative scores of those who did play will appear a little lower than they should be. In turn, Woods' relative scores will not stand out from the pack as much as they should. Indeed, when the simpler method of calculating the average score of a subset of tournament participants is used, Tiger Woods appears to be 2.6 strokes per round below the average of the top 100 money winners. With the adjustment based on the fixed-effects regression, Woods is more than 2.8 strokes better than the average.

[7]The leader in the number of tournaments entered was Mark Brooks who played in 34 of the 47 tournaments used. Gene Sauers played in only 11 tournaments because he had fallen off of the PGA TOUR priority list until he qualified for and won the Air Canada Championship halfway through the season.

above where Woods also leads at 2.83 strokes per round below *the average* of the other top 100 golfers. Although the unadjusted scoring average and the relative score average that I calculate are highly correlated with a simple correlation coefficient of .945, the correlation coefficient between VARIANCE and RELVAR falls off to .566, and falls even further to .457 for the skewness measures. Using the adjusted relative scores, variances, and measures of skewness does perform better in the regressions as illustrated in the next section.

Table 5.3 lists the five best golfers in each of the three statistical categories, along with their ranking in earnings per tournament. All the top leaders in having low RELMEAN scores relative to the competition are

Table 5.3 Top five lists.

Golfer	RELMEAN,	rank	RELVAR,	rank	RELSKEW,	rank	EARNPER rank
Tiger Woods	-2.826	1					1
Retief Goosen	-1.655	2					3
Vijay Singh	-1.475	3					5
Phil Mickelson	-1.461	4					4
Ernie Els	-1.454	5					2
Bob Burns			6.580	1			59
Jerry Kelly			5.548	2			10
Rory Sabbatini			5.273	3			57
Brandt Jobe			4.852	4			73
Steve Lowery			4.730	5			32
Tim Petrovic					-0.630	1	92
Billy Mayfair					-0.618	2	84
Craig Parry					-0.601	3	29
Lee Janzen					-0.550	4	62
Craig Perks					-0.548	5	38

familiar names, and they all are ranked highly in earnings per tournament. By contrast, being a leader in RELVAR or RELSKEW assures neither a high ranking in EARNPER, nor name recognition. However, by the table alone, one cannot rule out the possibility that these statistical "leaders" are not ranked highly in EARNPER because they are particularly bad in their RELMEAN scores. Such listings are interesting to golf fans and suggestive of the final results, however, the results of the multiple regression analysis reported in the next section allow us to simulate more powerfully the effects of improving on RELMEAN, RELVAR, or RELSKEW over the sample averages.

5.3 Results

Table 5.4 reports the results of O.L.S. regressions of EARNPER and LNEARNPER on alternative measures of mean performance, variance, and skewness. The table clearly illustrates the contention that using the distribution of relative scores rather than raw scores is worthwhile. Columns 1 and 3 use the averages, variances, and skewness measures from the actual scores and can be compared to columns 2 and 4 which use the averages, variances, and skewness measures from the relative scores.

The theoretical expectations are that golfers with lower relative scores will win more money so the sign on the estimated coefficient of SCOREAVE or RELMEAN should be negative. If having a large variance in scores is conducive to having big paydays occasionally, then the coefficient of the variance measure should be positive and significant. Finally, if having a large negative skewness to one's distribution allows one to really "go low," only rarely, but perhaps often enough to get a few very large paychecks, then the coefficient of SKEWNESS or RELSKEW should be negative and significant.

First, consider columns 1 and 2. In column 1 only SCOREAVE is significant and the adjusted R-square is 0.61. Meanwhile, in column 2, use of the relative scores improves the fit as indicated in the increase of the adjusted R-square to 0.77. Additionally, all three independent variables are statistically significant in the expected direction. A lower relative mean score increases earnings per tournament as does higher variance and a

Table 5.4 Regression coefficients (t-statistics).

Dependent Variable	(1) EARNPER	(2) EARNPER	(3) LNEARNPER	(4) LNEARNPER
Constant	4,853,000 (12.48)	10,075 (1.42)	69.39 (15.87)	10.15 (135.77)
SCOREAVE	- 67,881 (- 12.26)		- 0.831 (- 13.35)	
RELMEAN		- 66,507 (-18.25)		- 0.808 (- 21.03)
VARIANCE	847 (0.42)		0.026 (1.16)	
RELVAR		20,427 (7.86)		0.270 (9.87)
SKEWNESS	3,028 (0.49)		0.086 (1.25)	
RELSKEW		- 17,757 (- 4.22)		- 0.194 (- 4.38)
adjusted R^2	0.610	0.770	0.645	0.817

larger negative skewness. The economic meaning of the size of the coefficient is easily explained. Both columns 1 and 2 tell the same story, namely, that a decrease in strokes per round of one will lead to an increase of earnings per tournament in the neighborhood of $67,000.

In columns 3 and 4, the dependent variable is the natural logarithm of the earnings per tournament. By using the natural logarithm, the coefficient estimates indicate percentage changes.[8] Therefore, the numerical value of -0.831 for the variable SCOREAVE in column 3 means that a one stroke increase in score (per round) would lead to a decrease in earnings per tournament of about 83%. Since one stoke per round would translate to four strokes per tournament, such a sizeable decrease in earnings is an appropriate estimate. The effect of RELMEAN rounds to a similar 81% decrease for each additional stroke per round.

A comparison of columns 3 and 4 indicates that the adjusted measures of mean, variance and skewness outperform the unadjusted measures. Furthermore, none of the inferences stemming from columns 1 and 2 are changed. That is, in column 3, only SCOREAVE is significant and the adjusted R-square is only 0.65. By contrast, in column 4, using the relative scores instead of the absolute scores improves the fit as indicated by an increase in the adjusted R-square to 0.82. Meanwhile, the coefficients of RELMEAN, RELVAR, and RELSKEW are all significant and in the expected direction. A one unit increase in the variance of a golfer's scores leads to a 27% increase in earnings per tournament. Meanwhile, a one unit decrease in skewness, making the distribution more skewed to the left, leads to a 19.4% increase in earnings per tournament. It looks like consistency and heroics can each be a factor leading to success.

It is possible to get a feel for the relative importance of mean, variance, skewness by calculating some simple simulations using the results of the fourth column of Table 5.4. The "average" golfer with the sample average levels of RELMEAN, RELVAR, and RELSKEW, will earn the average logarithm of earnings per tournament of 10.82 which translates to $50,172 and would rank the 42nd highest in our sample. Now, consider holding RELVAR and RELSKEW constant and decrease RELMEAN to the leader's level of -2.826, a decrease of over 2.8 strokes. Multiplying this decrease by the coefficient of RELMEAN (equal to -0.808) the resulting logarithm of earnings per tournament increases to over 13 and the implied

[8]Use of the natural logarithm also corrects for a statistical problem known as heteroskedasticity that usually exists in the estimation of earnings functions, and is a common statistical device. Separate tests not reported here indicate that the problem exists in columns 1 and 2 and is solved in columns 3 and 4.

earnings per tournament would go up to $491,545 which would rank highest in the sample. This means that having the lowest average scores and only average variance and skewness are enough to lead the pack in earnings per tournament. In this sense, consistency surely pays.

Second, hold RELMEAN and RELSKEW constant and increase RELVAR to the level of the leader in that category. This experiment would increase earnings per tournament to $141,173 and the rank to fifth best. In this sense, heroics matters. A golfer with only average mean scores can climb into the top five in earnings per tournament if that average is comprised of a relatively large number of terrible performances balanced by a large number of superlative performances.

Finally, doing the same for RELSKEW while holding RELMEAN and RELVAR constant would only raise earnings per tournament to $60,828, and the related rank to 37th place. With average mean scores and variance, a distribution of scores skewed to the negative helps, but only slightly. These experiments indicate the relative importance of consistency in the sense of low mean scores versus heroics as measured by the variance and skewness of the distribution of a golfer's scores.

5.4 Discussion

The tournaments compensation model has been used to explain the large increases in salary that accompany promotions up the corporate ladder. This paper points out and verifies the possibility of a tendency to reward one-time, flash-in-the-pan performance in the tournaments model. At issue is whether promotion "tournaments" reward consistent performance in many "competitions," (for example a series of stellar yearly performance reviews) or heroic, extraordinary, recent performance (for example winning "salesman-of-the-year" recognition). The issue has both normative and positive aspects concerning the balance between the types of performance that salary structures do reward, and the types they should reward. The issue seems worthy of more close and careful consideration.

On a narrower front pertaining to golf, previous work has shown that golfer's yearly earnings and/or earnings per tournament are related to several areas of skill that are measured by the professional golf associations

[Shmanske, 1992, 2000, 2004a; Moy and Liaw, 1998; Nero, 2001; Rishe, 2001]. Longer, more accurate drives, accurate approach shots, and putting skills are all significant variables in regression equations for earnings for both men and women professional golfers. These previous results are actually a type of reduced form. The golfers use their skills to shoot low scores in one stage, and these low scores lead to high earnings in a second stage. The results cited above skip the intermediate step of scoring to directly find the effect of skills on earnings.

Scully [2002] has argued that more attention should be paid to the actual two stage structural model. This chapter gives additional impetus to this research imperative. Skills do not simply produce low mean scores, they produce a distribution of scores in which higher order moments of the distribution, like variance and skewness, also matter. Thus, there is much more to learn about the relationship between skills and earnings than is captured in the papers referenced in the preceding paragraph. Perhaps some skills, like driving distance, have little variation from tournament to tournament and lead to consistently lower scores, and thus higher earnings. But other skills, like driving accuracy or putting, might have large variances from tournament to tournament, thus leading to a large variance or skewness in the golfer's performance from tournament to tournament, with measurable consequences for the golfer's earnings. This chapter has established one imperative, namely, use of the year long scoring average, or of the tournament by tournament distribution of raw scores will not work because the raw scores are not comparable from tournament to tournament. Indeed, adjustments to the data like those described in this chapter have become the standard in research of this type.

This all begs the question of where variance and skewness of the golfer's relative scores come from. Do some golfers naturally have more variance and skewness in their scoring distributions, perhaps due to their style of play? Are variance and skewness of scores systematically related to the different skills such as driving distance or putting prowess? Or, possibly, do variance and skewness in scores simply represent the consequence of true random variation built into the nature of the game? To address these questions head on requires an examination of the microdata on golfers' skills measured at the level of the tournament rather than the season long averages which have been used in the previously referenced studies of earnings and skills. Thus, the mean, variance, and skewness of a golfer's relative scores, theoretically, are functions of the mean, variance, and skewness of the golfer's individual skills. The next chapter describes the collection and examination of this tournament level data. Read on.

Chapter 6

Skills, Performance, and Earnings in the Tournament Compensation Model: Evidence from PGA TOUR Microdata[1]

There were two important lessons in the previous chapter and both will be applied at a deeper level here. First, it was important to refine the year-long scoring average by looking at the individual tournament scores and adjusting them for the difficulty of the tournament course and the strength of the competition. Second, once the adjusted scores on a tournament-by-tournament basis were in hand, aspects of their distribution other than simply the average could be measured. This proved important because both the variance of the adjusted scores and their skewness turn out to be significant determinants of the golfer's earnings. This chapter applies these lessons to the underlying skills that produce the distribution of the scores in the first place. A little background may be in order.

In Shmanske [1992] I was one of the first to estimate a professional golfer's earnings function using regression analysis. That started a trend as several others followed suit using essentially the same model while making minor changes and additions.[2] Typically, these studies use year-

[1]The material in this chapter is drawn from Shmanske, S. (2008). Skills, Performance, and Earnings in the Tournament Compensation Model: Evidence from PGA TOUR Microdata, *Journal of Sports Economics*, 9(6), pp. 644-62.

[2]See Moy and Liaw [1998]; Nero [2001]; Shmanske [2000], [2004a]; Rishe [2001]; and Alexander and Kern [2005].

long average measures of skills, such as driving distance and putting proficiency, as explanatory variables, and earnings, earnings per tournament, or their logarithmic transformations as the dependent variable. These studies are called "reduced-form" equations because they combine in one step what is actually a two-step process. Skills do not directly produce earnings. Rather, in one step, the skills allow the golfers to shoot low scores in competitions. In a second step, it is the rank-order placement in the competitions that nonlinearly produces the earnings.

The results of these regressions are robust. Driving distance, putting proficiency, and accuracy with approach shots are almost always statistically and economically significant in the theoretically predicted direction. Driving accuracy and proficiency at recovering from sand bunker hazards are sometimes important. Other added explanatory variables such as age, age squared, experience, gender, or the number of tournaments entered, sometimes are significant but do not overturn the basic results measured for the skills. These models typically explain between 30 and 70 percent of the professional golfer's tournament earnings on the PGA TOUR.

A potential improvement was suggested by Scully [2002] who argued that a two-stage structural model should be examined. In a two-stage model earnings are a function of scores which are first a function of skills. More recently, Callan and Thomas [2007] have estimated a three-stage model in which skills produce scores which produce tournament ranks which produce earnings. These efforts led to only a modest improvement. Callan and Thomas change none of the basic inferences about which skills are important and offer only a slight improvement in explanatory power.

This brief background sets the stage to apply the first lesson referred to above. Indeed, one of the motivations for the research in this chapter is the possibility that the measurement protocol for the data, namely, year-long averages, has sabotaged the estimation of previous models. The problem with year-long averages is that they can lead to distorted measures of the underlying skills. For example, consider the effects of elevation and weather on the measurement of driving distance. Thin atmospheres at golf tournaments in Colorado and Lake Tahoe, allow the golf ball to fly roughly ten percent farther than usual. Meanwhile, tournaments played at sea level and/or in rainy conditions encourage neither long ball flight nor long amounts of roll once the ball hits the ground, thus leading to short drives. A golfer who plays in Colorado but not at Pebble Beach will have his driving distance overstated compared to his true level of skill, while the opposite will be true for a golfer competing at sea level but not in the mountains.

The same measurement error can exist for each of the golfer's skills, as for scoring in general. For example, consider driving accuracy which is the percentage of a golfer's tee shots that end up in the fairway. Since the fairways are different widths for each course, and since the penalty for driving off the fairway differs for each course, depending on whether off the fairway means in the woods, in a water hazard, in very tall grass, or simply in marginally taller grass, the measurement of accuracy on one course is not directly comparable to that at any other course. Yet, until recently, the PGA TOUR's measurement of year-long averages treats them as directly comparable.

Approach shot accuracy is measured as the percentage of greens that the golfer reaches in "regulation," that is, in two shots less than par for the hole. The size and firmness of the greens and the overall length of the course will directly affect the measurement of approach shot accuracy. Overall, PGA TOUR golfers hit about two-thirds of the greens in regulation. On some courses, obtaining this average would correspond to a stellar performance, but on others, "only" two-thirds would be considered dismal.

Putting prowess is measured by the average number of putts taken only on those greens reached in regulation. This controls somewhat for golfers who take fewer putts because they more often miss the green with the approach shot but then need only one putt after a short chip or pitch shot leaves them very close to the hole. The size and contour of the greens, which differ across tournament venues, obviously influence this measurement of the putting skill. Large greens, for example, translate into more greens being reached in regulation, but resulting in a larger number of long putts and a higher number of putts per green.

Sand bunker skill is measured by the percentage of times a golfer finishes a hole in one or two shots from a greenside bunker. Therefore, the number, position, size, and depth of the sand bunkers will cause systematic differences in the measurement of the sand bunker recovery skill.

In addition to each of the skills, the overall performance of the golfer, as measured by the average scores, can also be distorted. Clearly, the overall difficulty of the golf course and the conditions, including wind and weather, differ from week to week making the scoring average measure not directly comparable from tournament to tournament. It is remarkable that Tiger Woods competed in only the toughest tournaments with the highest average scores overall, and skipped tournaments on easier golf courses where everyone shoots under par, and *still* had the lowest scoring average of all professionals in the 2006 data used in this chapter. Precise measures of Woods's dominance will be highlighted below.

Fortunately, if data can be collected for golfer skills on a tournament-by-tournament basis, then each of the skill measurements can be adjusted by reference to the average level of skill shown in the tournament. For example, if everyone in the tournament is hitting 300-yard drives because of high altitude, then any particular golfer's 300-yard drive is not such a remarkable achievement as it first seems. The details of the adjustment are described below. A comparison of a typical reduced-form earnings regression with the adjusted and unadjusted skill measures demonstrates the effect of the improved measurement protocol adopted in this research.

In addition to refining the measurements of the average levels of skill and the scoring averages, the capturing of the tournament-based data allows the application of the second lesson referred to in the opening paragraph. Namely, by capturing multiple observations (one for each tournament) of each skill and of overall scores, other aspects of the distributions of skills and scores for each golfer can be examined. When constrained to use year-end averages, then average levels are the only data that one has. But if a distribution of measures for each of the skills is observed then parameters other than the means of these distributions, such as the variance and skewness, can be calculated for each of the skills.

Chapter 5 has shown that yearly winnings per tournament are a function not only of (adjusted) scoring average, but also of the variance and the skewness of the distribution of (adjusted) scores that the golfer achieved over the course of the year. Considering a two-stage structural model again, previous attempts to link earnings (or earnings per tournament) to scoring *average* have been incompletely specified since earnings are a function of mean, variance, and skewness of the distributions. And if the second stage (earnings as a function of performance) has been underspecified, so has the first stage (performance as a function of skills). Indeed, this research attempts to find out where the variance and skewness in scoring come from. For example, is the variance or skewness in the scoring distribution due to variance or skewness in the skill of driving distance, variance or skewness in the putting distribution, both, neither, or something else?

The collection of the raw data and the adjustments that are made to bring about better comparability across tournaments are discussed in Section 6.1. The model specifications to be estimated are presented in Section 6.2. Section 6.3 reports the results and a final section summarizes.

6.1 Data Measurement and Adjustment

The PGA Tour reports on its website, on a year-to-date basis, a variety of statistical performance measures by individual golfer.[3] Although the PGA TOUR would not share its microdata,[4] it was possible to track the performances of a pre-chosen set of golfers, "backing out" the most recent weekly performance statistics from the change, if any, in the year-to-date statistics. For example, suppose that the driving distance of a golfer after his first tournament of the year is reported as 280 yards. After the second tournament, his year-to-date driving distance is now listed as 290 yards. Then, in order for this measure to have increased by 10 yards it must have been that the driving distance in the second tournament was 300 yards. The two individual observations of 280 yards and 300 yards would average to 290 yards as reported.

The calculation can be generalized. Let X_t denote the year-to-date average after t tournaments, X_{t+1} the average after t + 1 events, and A_{t+1} the actual performance in the t + 1st tournament. Then the relationship among these variables is given by:

[3]Weekly visits to www.pgatour.com were made during the 2006 golfing season.

[4]To their credit the PGA TOUR does now give access to academic researchers.

$$X_{t+1} = (tX_t + A_{t+1})/(t+1)\ . \tag{6.1}$$

Solving this for A_{t+1} yields:

$$A_{t+1} = (t+1)X_{t+1} - tX_t\ . \tag{6.2}$$

This tedious process was performed to obtain the week-by-week performances for each tournament entered in 2006 by each of the top 100 money winners of 2005.[5] Forty-six of the forty-eight official tournaments are used in the data.[6] This led to 2,360 individual observations. The busiest golfers entered 32 tournaments, while at the low end, Loren Roberts played in only four tournaments because he turned 50 in 2005 and played mostly on the Senior Tour in 2006.

For each of these 2,360 observations six statistics were tracked: The score per 18 holes measured in strokes, SCORE; the driving distance measured in yards, DRIVDIST; the driving accuracy measured as the percentage of drives ending in the fairway, DRIVACC; approach shot accuracy measured as the percentage of greens reached in regulation, GIR; putting proficiency measured as the number of putts taken per green reached in regulation, PUTTPER; and sand bunker skills measured as the percentage of times two or fewer strokes are taken to finish a hole from a greenside bunker, SANDSAVE. Summary statistics for these, their transformations (explained below), and the other variables used appear in Table 6.1.

As discussed in the introduction, the observations for each of these variables are not directly comparable across tournaments, therefore, adjustment factors for each course are sought. To obtain course adjustment

[5]The choice of 100 is arbitrary. Other obvious possibilities would have been 125, the number of golfers with fully exempt status, 144, the number allowed to enter a typical tournament, or perhaps even every golfer who played in a tournament or made a cut in a tournament. Larger samples increase the workload for little benefit because 100 golfers leaves ample degrees of freedom to uncover the significant relationships.

[6]Two tournaments, The INTERNATIONAL and the Accenture Match Play Championship, are omitted because they use alternative scoring protocols in which golfers play different numbers of holes and not every shot counts.

Table 6.1 Summary statistics.

Variable	Mean	Std. Dev.	Minimum	Maximum	N
SCORE	71.46	2.32	65	82	2360
DRIVDIST	289.09	13.90	184.4	353.2	2360
DRIVACC	0.636	0.120	0	1	2360
GIR	0.655	0.083	0.38	1	2360
PUTTPER	1.777	0.090	1.403	2.133	2360
SANDSAVE	0.496	0.207	0	1	2360
ASCORE	3.024	1.874	-3.793	11.996	2360
ADRIVDIST	-18.052	10.347	-120.241	32.324	2360
ADRIVACC	0.008	0.095	-0.664	0.605	2360
AGIR	-0.102	0.068	-0.384	0.409	2360
APUTTPER	0.031	0.082	-0.289	0.346	2360
ASANDSAVE	-0.050	0.201	-0.675	0.537	2360
DOLLARS	1,640,509	1,436,855	13,432	9,941,563	100
DOLLARS/N	71,258	80,628	2,239	662,771	100
LnDOLLARS/N	10.8	0.874	7.714	13.404	100
SCOREMEAN	70.82	0.790	68.11	73.71	100
ASCOREMEAN	70.82	0.925	67.85	74.04	100
ASCOREVAR	2.995	1.252	0.945	8.242	100
ASCORESKEW	0.405	0.562	-1.015	1.896	100
DRIVDISTMEAN	289.15	7.835	265.8	307.1	100
ADRIVDISTMEAN	-17.94	7.714	-35.15	0.529	100
ADRIVDISTVAR	55.54	82.87	11.94	842.34	100
ADRIVDISTSKEW	-0.145	0.702	-3.59	1.38	100
DRIVACCMEAN	0.635	0.055	0.505	0.784	100
ADRIVACCMEAN	0.008	0.521	-0.121	0.153	100
ADRIVACCVAR	0.007	0.005	0.002	0.035	100
ADRIVACCSKEW	0.164	0.528	-1.531	1.591	100
GIRMEAN	0.655	0.029	0.576	0.756	100
AGIRMEAN	-0.099	0.033	-0.173	0.042	100
AGIRVAR	0.004	0.003	0.002	0.035	100
AGIRSKEW	0.064	0.534	-0.915	1.775	100
PUTTPERMEAN	1.777	0.025	1.712	1.849	100
APUTTPERMEAN	0.030	0.029	-0.047	0.115	100
APUTTPERVAR	0.006	0.003	0.001	0.027	100
APUTTPERSKEW	-0.129	0.582	-1.891	1.292	100
SANDSAVEMEAN	0.499	0.056	0.344	0.636	100
ASANDSAVEMEAN	-0.048	0.055	-0.211	0.096	100
ASANDSAVEVAR	0.041	0.018	0.012	0.121	100
ASANDSAVESKEW	0.110	0.546	-1.798	1.911	100

factors for each of the variables, six two-dimensional fixed effects equations were fitted with ordinary least squares. The equations take the form:

$$X = B_C C + B_G G + E \qquad 1 \leq i \leq 46 \text{ and } 1 \leq j \leq 100, \tag{6.3}$$

where X is the vector containing the dependent variable, SCORE, DRIVDIST, etc., with element x_{ij} capturing the performance observed on the ith course by the jth golfer, B_C is the vector of 46 coefficients for the course adjustment factors, C is the corresponding matrix of dummy variables, one for each course, B_G is the vector of 99 coefficients (one for each golfer with Tiger Woods omitted to avoid singularity in the estimation process) to control for the fact that golfers with different sets of skills play in different tournaments, G is the matrix of dummy variables, one for each golfer, and E, with individual element, e_{ij}, is the vector of error terms. In Eq. (6.3) the concern is with the estimates of B_C which will be used to adjust the raw data, but it is desirable to include G to control for the golfers actually playing in the tournament as the following numerical thought experiment will illustrate.

Suppose three golfers of average driving distance ability record driving distances of 310, 300, and 290 yards in a given tournament. If these are the only three golfers in the tournament, the course adjustment factor would be 300, and the golfer's "adjusted" or "relative" driving distances would be 10, 0, and -10, for that tournament. In another tournament, Tiger Woods (who hits drives 10 yards farther than average, ceteris paribus) and two average golfers record drives of 300, 280, and 270 yards respectively. If the golfer fixed effects in $B_G G$ were omitted from Eq. (6.3), then the adjustment factor for this course would be 283.33 and the adjusted driving distances would be 16.67, -3.33, and -13.33. However, this does not do Tiger, or the other golfers in the tournament with Tiger, justice. With $B_G G$ included, 10 yards of Tiger's 300 will be attributed to Tiger in B_G, leaving only 290 to be averaged in to get the course adjustment factor, which would now equal 280. The adjusted driving distances of the three golfers would now be 20, 0, and -10, that is, 3.33 yards longer than without the golfer fixed effects in $B_G G$.

The adjustments are made to all six of the observed variables by subtracting the corresponding coefficient in B_C from each of the observations. For example, if x_{ij} corresponds to the SCORE for the jth golfer on the ith course, then the ith element of B_C from the SCORE equation is subtracted from x_{ij} to yield the adjusted or relative score for the jth golfer on the ith course, and will be denoted as ASCORE. Similarly, ADRIVDIST, ADRIVACC, AGIR, APUTTPER, and ASANDSAVE will signify the adjusted or relative levels of the individual skill variables. Note once more that this adjustment process required the exclusion of the dummy variable for one of the golfers, chosen arbitrarily to be Tiger Woods. Consequently, the results are all couched in terms of differences between Tiger's averages and the observation in question. For example, in Table 6.1 examine the ADRIVDIST variable. The average of all 2,360 adjusted drives recorded is 18.052 yards shorter than Tiger's average. The shortest ADRIVDIST recorded is more than 120 yards shorter than Tiger's average. This outlying observation must be owing to a particularly short drive that perhaps hit a tree and rebounded backwards, something with which we are all probably familiar.

This process yields 2,360 individual adjusted observations on each of six variables. For each of the six, the observations will be grouped by golfer to yield 100 distributions of each of the skills. Finally, for each distribution, the mean, variance, and skewness can be calculated in straightforward manner. This yields 100 observations, one for each golfer, which will be used in the models described in the next section. Each observation will have: yearly earnings, DOLLARS; the number of tournaments, N; the earnings per tournament, DOLLARS/N, and its natural logarithm, LnDOLLARS/N; the unadjusted scoring average, SCOREMEAN; the mean, variance, and skewness of the adjusted scoring average distributions, ASCOREMEAN, ASCOREVAR, ASCORESKEW; and the mean, variance, and skewness of each of the adjusted skill measures, ADRIVDISTMEAN, ADRIVDISTVAR, ADRIVDISTSKEW; and so on.

For just one illustration of the difference made by the measurement protocol adopted here, consider Tiger Woods and the scoring average variable. With the unadjusted statistics Tiger's scoring average is 68.11 strokes per 18 holes. The average scoring average of the 100 golfers in the sample is 70.82 strokes per 18 holes. Woods is better than the average by about 2.71 strokes per round. But this underestimates Tiger's dominance. When the adjusted or relative scoring averages are used, Tiger is better than the average by 3.02 strokes which translates to a scoring average of 67.8 per 18 holes. The difference between the two measurement protocols amounts to more than one stroke per tournament. An alternative point of comparison is with the second best golfer of 2006, Jim Furyk. Woods is better than Furyk by 0.75 strokes per round (a whopping three shots per tournament) in the unadjusted data and by 0.85 strokes per round after the adjustments are made.

6.2 Model Specification

To facilitate comparisons with the received literature, the first model to be estimated is a bare-bones reduced-form model of the following form:

$$\text{LnDOLLARS/N} = \text{BZ} + \text{E}, \qquad (6.4)$$

where the dependent variable is the natural logarithm[7] of the yearly average earnings per tournament, Z is the vector of the constant term and five skills measures, B is the vector of unknown coefficients to be estimated and E contains the error terms. Equation (6.4) will be estimated with ordinary least squares for the unadjusted data (referred to as Eq. (6.4a)) and re-estimated using the adjusted data (referred to as Eq. (6.4b)). The results appear in Table 6.2. As is easily seen, the precision of the estimates and the overall explanatory power both improve.

To help describe the two-stage structural model consider Fig. 6.1. The top panel in the figure corresponds to the simple reduced-form model in Eq. (6.4) and estimated in Table 6.2. Now move to the middle panel. The simple structural model described in the second panel breaks the relationship between skills and earnings into two steps. First, skills produce low scores in tournaments, and, second, those tournament

[7]As usual there is heteroskedasticity when earnings levels are used, but it disappears when the logarithmic transformation is applied.

Table 6.2 Parameter estimates and (t-statistics) for Eqs. (6.4a) and (6.4b).

Equation Dependent Variable	(6.4a) LnDOLLARS/N	(6.4b) LnDOLLARS/N
constant	17.18** (2.42)	11.26*** (79.05)
DRIVDISTMEAN	0.0264** (2.17)	
ADRIVDISTMEAN		0.0326*** (3.31)
DRIVACCMEAN	1.866 (1.01)	
ADRIVACCMEAN		3.402** (2.16)
GIRMEAN	10.34*** (3.71)	
AGIRMEAN		9.333*** (4.51)
PUTTPERMEAN	-13.22*** (-4.53)	
APUTTPERMEAN		-14.30*** (-6.94)
SANDSAVEMEAN	3.026** (2.23)	
ASANDSAVEMEAN		2.356** (2.19)
adjusted R^2	0.362	0.615
N	100	100

Note: *, **, *** indicate statistical significance at the .10, .05, and .01 levels respectively.

performances translate into earnings. These two steps are captured in Eqs. (6.5) and (6.6):

$$\text{Performance} = BZ + E \qquad (6.5)$$

$$\text{LnDOLLARS/N} = b_0 + b_1\text{Performance} + E. \qquad (6.6)$$

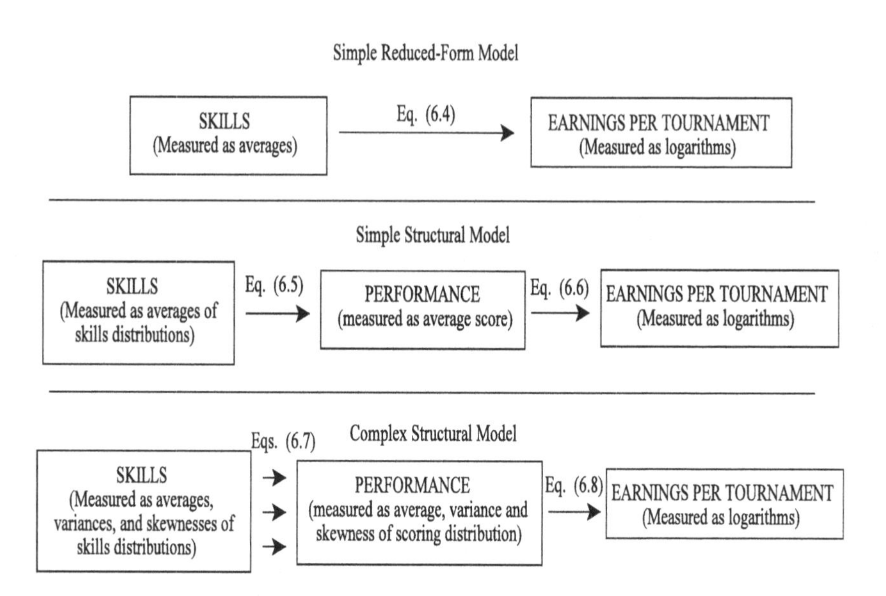

Fig. 6.1 Reduced-form and structural models.

In Eqs. (6.5) and (6.6) B, Z, E, and LnDOLLARS/N are as previously defined, and performance will be measured by the scoring average. Also as above, these equations will be estimated with the unadjusted data for skills and scoring average, referred to as Eqs. (6.5a) and (6.6a) and re-estimated with the adjusted data as Eqs. (6.5b) and (6.6b). The extent to which the received practice of using year-long averages in the measurement of skills and performance has hampered the ability to resolve the structural coefficients in Eqs. (6.5) and (6.6) will be illustrated by the comparison of the results with and without the data adjustments. These results appear in Table 6.3. Although the magnitudes of the coefficients are comparable, the precision of the estimates as measured by the t-statistics and the adjusted R^2 improve as we move from Eqs. (6.5a) and (6.6a) to Eqs. (6.5b) and (6.6b).

The third panel in Fig. 6.1 captures the second improvement that becomes possible with the collection of tournament-by-tournament data. Namely, the measurement of performance and skills is no longer limited

Table 6.3 Parameter estimates and (t-statistics) for Eqs. (6.5) and (6.6).

Equation	(6.5a)	(6.5b)	(6.6a)	(6.6b)
Dependent Variable	SCOREMEAN	ASCOREMEAN	LnDOLLARS/N	LnDOLLARS/N
constant	66.58*** (12.02)	70.22*** (587.17)	82.36*** (25.56)	72.77*** (28.45)
SCOREMEAN			-1.010*** (-22.21)	
ASCOREMEAN				-0.875*** (-24.23)
DRIVDISTMEAN	-0.02251** (-2.37)			
ADRIVDISTMEAN		-0.03213*** (-3.88)		
DRIVACCMEAN	-2.047 (-1.42)			
ADRIVACCMEAN		-3.500*** (-2.64)		
GIRMEAN	-12.605*** (-5.80)			
AGIRMEAN		-12.67*** (-7.29)		
PUTTPERMEAN	12.392*** (5.44)			
APUTTPERMEAN		15.72*** (9.08)		
SANDSAVEMEAN	-3.425*** (-3.24)			
ASANDSAVEMEAN		-2.474*** (-2.73)		
adjusted R^2	0.525	0.757	0.833	0.856
N	100	100	100	100

Notes: *, **, *** indicate statistical significance at the .10, .05, and .01 levels respectively.

to looking only at the average of the distributions. Thus, the following set of equations will be estimated in a two-level structural model:

$$\text{ASCOREMEAN} = \text{B'Z'} + \text{E} \quad (6.7.1)$$
$$\text{ASCOREVAR} = \text{B'Z'} + \text{E} \quad (6.7.2)$$
$$\text{ASCORESKEW} = \text{B'Z'} + \text{E} \quad (6.7.3)$$

$$\text{LnDOLLARS/N} = b_0 + b_1\text{ASCOREMEAN} + b_2\text{ASCOREVAR} + b_3\text{ASCORESKEW} + \text{E}, \quad (6.8)$$

where Z' captures the characteristics of the skills distributions including the variances and skewnesses of those distributions along with the averages that were already included in Z as used in Eqs. (6.4) and (6.5). Equations (6.7.1-6.7.3) capture the first stage in the model, namely, skills produce performance. But three aspects of the performance distribution are important as verified in Eq. (6.8) which captures the second stage of the structural model. The means, averages, and skewnesses of the adjusted scoring distributions each independently influence the earnings per tournament. Equations (6.7.1-6.7.3) will indicate which aspects of the skills distributions lead to which aspects of the scoring distribution. The equations will be estimated only with the adjusted measures. These are the main results of the chapter. They are listed in Table 6.4 and will be discussed below along with the rest of the results.

Two more equations are estimated as a type of double-check on the specifications discussed above, and the results are listed in Table 6.5. The first of these is:

$$\text{LnDOLLARS/N} = \text{B'Z'} + \text{E}, \quad (6.9)$$

which is a reduced-form specification similar to received models, but which adds the extra variances and skewnesses to Eq. (6.4), which had only considered the means. Equation (6.9) can be directly compared with Eq. (6.4b) to show the improvement made by considering the higher order parameters of the observed skill distributions.

Table 6.4 Parameter estimates and (t-statistics) for Eqs. (6.7) and (6.8).

Equation	(6.7.1)	(6.7.2)	(6.7.3)	(6.8)
Dep. Var.	ASCOREMEAN	ASCOREVAR	ASCORESKEW	LnDOLLARS/N
constant	69.69***	1.651***	0.0644	75.24***
	(445.85)	(3.10)	(0.25)	(35.24)
ASCOREMEAN				-0.9157***
				(-30.18)
ASCOREVAR				0.1618***
				(6.88)
ASCORESKEW				-0.1815***
				(-3.55)
ADRIVDISTMEAN	-0.01108	0.0569**	0.01608	
	(-1.54)	(2.33)	(1.39)	
ADRIVDISTVAR	0.000439	0.000703	-0.000018	
	(0.87)	(0.41)	(-0.02)	
ADRIVDISTSKEW	0.02209	0.2127	-0.0365	
	(0.36)	(1.01)	(-0.36)	
ADRIVACCMEAN	-0.844	7.255*	3.295*	
	(-0.70)	(1.77)	(1.69)	
ADRIVACCVAR	19.65**	91.55***	5.493	
	(2.04)	(2.79)	(0.35)	
ADRIVACCSKEW	0.0675	0.3332	0.03587	
	(0.88)	(1.27)	(0.29)	
AGIRMEAN	-17.24***	-15.80***	-4.026	
	(-10.91)	(-2.94)	(-1.58)	
AGIRVAR	72.14***	-18.92	46.54*	
	(4.35)	(-0.34)	(1.74)	
AGIRSKEW	-0.0654	0.1979	0.07296	
	(-0.91)	(0.81)	(0.63)	
APUTTPERMEAN	16.67***	4.993	0.3859	
	(12.07)	(1.06)	(0.17)	
APUTTPERVAR	36.45***	-1.197	8.760	
	(2.61)`	(0.03)	(0.39)	
APUTTPERSKEW	-0.0435	-0.1491	0.1078	
	(-0.66)	(-0.66)	(1.01)	
ASANDSAVEMEAN	-2.500***	3.089	0.6876	
	(-3.29)	(1.19)	(0.56)	
ASANDSAVEVAR	-4.097*	10.32	-1.320	
	(-1.66)	(1.24)	(-0.33)	
ASANDSAVESKEW	-0.143*	-0.4511	0.1095	
	(-1.75)	(-1.62)	(0.83)	
adjusted R^2	0.858	0.100	0.00	0.903
N	100	100	100	100

Note: *, **, *** indicate statistical significance at the .10, .05, and .01 levels respectively.

Table 6.5 Parameter estimates and (t-statistics) for Eqs. (6.9) and (6.10).

Equation	(6.9)	(6.10)
Dep. Var.	LnDOLLARS/N	LnDOLLARS/N
constant	11.77***	74.73***
	(57.58)	(12.33)
ASCOREMEAN		-0.9073***
		(-10.43)
ASCOREVAR		0.1692***
		(6.21)
ASCOREKEW		-0.1495***
		(-2.60)
ADRIVDISTMEAN	0.0143	-0.00297
	(1.53)	(-0.49)
ADRIVDISTVAR	-0.00054	-0.0003
	(-0.81)	(-0.64)
ADRIVDISTSKEW	-0.01227	-0.0337
	(-0.15)	(-0.68)
ADRIVACCMEAN	1.699	0.1985
	(1.08)	(0.20)
ADRIVACCVAR	-8.599	-5.442
	(-0.68)	(-0.66)
ADRIVACCSKEW	-0.00382	0.0064
	(-0.04)	(0.10)
AGIRMEAN	13.37***	-0.1997
	(6.47)	(-0.10)
AGIRVAR	-86.43***	-10.81
	(-3.99)	(-0.73)
AGIRSKEW	0.03939	-0.04251
	(0.42)	(-0.74)
APUTTPERMEAN	-15.77***	-1.430
	(-8.73)	(-0.79)
APUTTPERVAR	-34.04*	0.5478
	(-1.87)	(0.05)
APUTTPERSKEW	-0.05813	-0.05622
	(-0.67)	(-1.06)
ASANDSAVEMEAN	2.115**	-0.5725
	(2.13)	(-0.88)
ASANDSAVEVAR	3.948	-1.713
	(1.23)	(-0.85)
ASANDSAVESKEW	-0.02411	-0.0615
	(-0.23)	(-0.90)
adjusted R^2	0.728	0.900
N	100	100

Note: *, **, *** indicate statistical significance at the .10, .05, and .01 levels respectively.

Finally, a "kitchen sink" model including all the regressors from the complex, two-stage model is estimated as:

$$\text{LnDOLLARS/N} = b_0 + b_1\text{ASCOREMEAN} + b_2\text{ASCOREVAR} + b_3\text{ASCORESKEW} + B'Z' + E. \quad (6.10)$$

If the parameters of the skill distributions have independent explanatory power for earnings, over and above that which affects the tournament performance as captured in the scoring distributions, then the parameters estimated in B' will be significant. There is no theory to suggest that this should be the case, and the coefficients in B' are not expected to be significant. However, such an equation might be appropriate if the dependent variable included earnings other than tournament purses. For example, the ability to hit prodigiously long drives might influence earnings indirectly through its effect on shooting low scores in tournaments, and directly because of its effect on product endorsement deals. A more sophisticated approach may be required to untangle the direct and indirect effects of skills on earnings in such a model. The model as construed in this chapter, and confirmed by the insignificant results in Eq. (6.10), is sequential rather than simultaneous, and ordinary least squares performed separately at each level is the appropriate estimation procedure.

6.3 Results

Consider Table 6.2 which shows the results of estimating a reduced-form model with the adjusted Eq. (6.4b) and unadjusted Eq. (6.4a) data. There is clear evidence that the data adjustments improve the regression in that the adjusted R^2 jumps from 0.362 to 0.615. On the three skills variables that consistently have been shown to be important, driving distance, greens in regulation, and putting, the improved measurements lead to lower standard errors but no change in inference. Driving distance, approach shot accuracy, and putting prowess are once again shown to have significant measurable effects on earnings. However, there is a change in inference with respect to driving accuracy, which is now significant with the adjusted data.

The magnitudes of the effects in Eq. (6.4b) are reasonable. The coefficient of the driving distance variable, 0.0326, means that an extra yard of distance on average will increase earnings by a little over three percent. An extra *percentage point* in approach shot accuracy will increase earnings by about 9.3 percent. To interpret the putting coefficient, -14.29, a little more work is required. Of course, the coefficient should be negative because fewer putts are better than more putts. Suppose a golfer holed out one more putt per round on a green reached in regulation, thus avoiding one extra short tap in putt. Since golfers reach about two-thirds of the greens in regulation, that is, about 12 of the 18 holes per round, we must divide the coefficient by 12 to get the effect. Thus, a golfer taking one fewer putt per round (four strokes per tournament) will increase earnings by about 119 percent. Given the nonlinear structure of the tournament payouts, this figure is very reasonable.

Table 6.3 presents the results for the simple structural model. All the coefficients are the expected sign, and the magnitudes are reasonable. By comparing Eq. (6.5b) to Eq. (6.5a) we again see the improvement made when the adjusted data is used. With the adjusted data, all five skills are statistically significant, whereas driving accuracy fails to reject the null with the unadjusted data, and the adjusted R^2 improves from 0.525 to 0.748. In Eq. (6.6) the use of the adjusted scoring average improves the standard errors and the adjusted R^2 only slightly, without changing the basic inference that low scores lead to higher earnings. The magnitude of the effect is that a one stroke improvement per round (four strokes per tournament) leads to about a doubling of earnings per tournament. The point estimate is 101 percent in Eq. (6.6a) and 88 percent in Eq. (6.6b).

Table 6.4 lists the results for the complex structural model in which higher order parameters of the skills and scoring distributions are used. First consider Eq. (6.8) which documents the improvement made when the variance and skewness of the adjusted scoring distributions are used to explain the logarithm of earnings per tournament. Equation (6.8) clearly shows that the shape of the distribution of the adjusted scores is important in addition to the simple mean of the distribution. Low scores are obviously rewarded, the point estimate of a 91.6 percent increase in earnings for a one stroke per round improvement reinforces the results from Table 6.3. But now, as in Chapter 5, it is obvious that variance and skewness also matter. Holding adjusted scoring average constant, a greater variance and/or a greater negative skewness in one's performances pays off. This is the expected result. Because of the nonlinear payout structure in the tournament compensation scheme, it is better to be farther below your average roughly half the time balanced by above average scores, that is, high variance, or to be way below your average only occasionally, that is, negative skewness, than it is to hit your average every time. By including variance and skewness, the explanatory power increases compared to Eq. (6.6b) as indicated by the jump in adjusted R^2 from 0.856 to 0.903.

The results of Eq. (6.8) confirm the results from the 2002 data used in Chapter 5. Mean, variance, and skewness of each golfer's distribution of scores determine the golfer's earnings per tournament. But where do mean, variance, and skewness come from? Equations (6.7.1-6.7.3) are estimated to answer this question. The results are enlightening. First, consider Eq. (6.7.1) in which the mean of the distribution of adjusted scores is the dependent variable. By adding higher order parameters of the skills distributions, Eq. (6.7.1) increases the adjusted R^2 from the corresponding equation in Table 6.3, Eq. (6.5b), from 0.748 to 0.858. This represents an almost 15 percent increase in the ability to explain the variation in average scores across golfers. Looking at the individual explanatory variables we see that the mean levels of accuracy with approach shots, putting prowess, and sand bunker skills are all still statistically significant with magnitudes comparable to those estimated in Eq. (6.5b). However, the mean levels of driving distance and accuracy are no longer significant. The increase in explanatory power comes mainly from several of the variance measures which are now significant. Variance in the golfer's performances with respect to the skills of driving accuracy, approach shot accuracy, and putting ability all lead to increased, that is worse, scores. Interestingly, higher variance in sand saves leads to lower scores. Meanwhile, the only inference about the skewness measures that is significant is the negative relationship between skewness in the distribution of sand saves and the average score per round. A positive skewness in sand saves, that is a few really good tournaments with respect to recovering from bunkers, will lower the average scoring for the season.

It is interesting how the variance in the skills measures matters for the average score. Except for sand saves, when variance matters, it is always positively related to scores; in golf, where lower scores are better, this is a bad thing. Evidently, the up and down variation around the average level of skill does not have a symmetric effect on scores. Consider variance in the skills of approach shot accuracy, measured by AGIRVAR, and driving accuracy, measured by ADRIVACCVAR. When a golfer hits more greens in regulation or more fairways than usual, it does not lower his score as much as his score is raised in those cases when a golfer hits fewer greens in regulation than usual or misses the fairway more often than usual. This makes sense. For example, when a golfer hits a green in regulation, he has a chance to save one stroke by holing out his first putt instead of taking two putts which is the modal case. On average, less than a stroke will be saved in this case. Alternatively, when a golfer misses the green in regulation it will usually cost the golfer and sometimes the cost might be multiple recovery strokes or penalty strokes depending on where the errant shot ended up. On relatively rare occasions, missing the green in regulation does not cost the golfer (chip-ins for birdie or up-and-downs for par are possible), but more often than not, the extra strokes caused by missing the green are more than the strokes saved by reaching the green in regulation.

Meanwhile, Eqs. (6.7.2-6.7.3) which model the variance and skewness of the scoring distributions are interesting in their lack of predictive power. Only four of the 15 variables are significant in the ASCOREVAR equation leading to an adjusted R^2 of 0.10, and only two are significant at even the ten percent level in the ASCORESKEW equation which appears to have no explanatory power at all. This result is consistent with the conclusion that variance and skewness of the scoring distributions are essentially random variations. Of course, the inability to model or predict random variation is not a flaw in the econometrics, rather, it is the expected result, if the variations around the mean of the scoring distributions do, in fact, represent true random variation. There is a little evidence that the variance of the scoring distributions is systematically related to the means of driving distance and greens in regulation, and the mean and variance of driving accuracy. In particular long drives and increased variance in driving accuracy each lead to increased variance in the scoring department. Meanwhile, the way to reduce variance in scoring is to hit more greens in regulation on average. However, most of the variance in scores remains unexplained.

Random variation is, perhaps, one of the things behind the fascination with sports in general, and the sport of golf in particular. Interestingly, however, because of the nonlinear payout structure in the tournaments compensation model, this random variation has economic consequences. Ultimately, it may be impossible to bring about extra effort by rewarding stellar performances without also rewarding luck and random variation.

Briefly moving to Table 6.5 and Eqs. (6.9) and (6.10), the results confirm and support several of the findings of this chapter. Equation (6.9) is a reduced-form model that includes both the improvements in the measurement of the data and the extra information about the distributions of the skills. As such, it is comparable to Eqs. (6.4a-6.4b). In Eq. (6.4a) only the mean levels of skill are used with the unadjusted skills data and the adjusted R^2 is only 0.362. The data improvement in Eq.(6.4b) led to an increase in explanatory power of the model to an adjusted R^2 of 0.614. Adding the variances and skewnesses of the skills distributions in Eq. (6.9) further increased the explanatory power to an adjusted R^2 of 0.727 for an increase of over 18 percent. With respect to the individual variables, mean and variance of both putting and approach shot accuracy, along with the mean of sand bunker recovery skills are factors which are statistically significant in the expected direction. Hitting more greens in regulation will raise income, but variance in this skill will lower income. Taking fewer putts on greens reached in regulation will raise income, but variance in putting prowess will lower income. Interestingly, the coefficient of the driving distance variable falls enough to remove it from the list of significant variables. Driving distance is significant in the structural model, and has usually been significant in the reduced-form models in the received literature, but in Eq. (6.9) driving distance has lost some of its luster.

Turning to Eq. (6.10) we find that there is no independent effect of the skills on earnings that shows up once the mean, variance, and skewness of the scoring distributions are accounted for. This supports the formulation as a separable two-level model in which skills affect scores at one level and scores produce earnings at another level.

6.4 Summary

This chapter estimated a structural model of tournament earnings by PGA TOUR golfers. The collection of data for each golfer at the tournament level allowed for a better calculation of the average level of skills exhibited by the golfers while at the same time allowing for the calculation of other parameters of the skills distributions such as variance and skewness. The chapter also estimated reduced-form models with and without the data adjustments to illustrate both the extra precision that the data adjustments make, and the extra precision that the addition of variance and skewness measures makes.

Adding variance and skewness of the scoring distributions to the earnings function is an important improvement over the earlier literature. Both variance and skewness are significant in the predicted directions. But the question of where variance and skewness come from remains unsettled. There is evidence that a small portion of the variance in the scoring distribution function can be related to parameters of the skills distribution functions, but the overwhelming majority of the variation across golfers in variance measures and all of the variation in the skewness parameter remain unexplained. This result is consistent with the view that skewness and variance in the distribution of scores simply represent true random variation. This type of uncertainty of outcome surely adds drama and pathos, and probably adds to the viewing and rooting pleasure of fans. But this random variation is also rewarded generously by the nonlinearity of the tournament payouts. While other research has shown that it is possible to bring forth extra effort with a tournament compensation scheme, it may be impossible to reward effort indirectly by rewarding performance without also rewarding luck at the same time.

Chapter 7

Gender Discrimination Revisited[1]

The preceding chapter showed how a micro-intensive look at the relationships among skills, scoring, and earnings in professional golf yielded more precise measurements of the actual numerical connections among them. Economically speaking, however, it is not necessarily the case that the extra precision obtained is worth the cost of the extra data manipulation. To push scientific frontiers forward, the extra cost is probably worth it, and in any case is probably necessary for editors to be sufficiently impressed to allow scientific publication. That said, there were not any eye-opening changes to the basic qualitative inferences about the relationships between the skills that golfers exhibit and the earnings they make. For example, driving the ball farther and straighter, hitting more greens, and making more putts, all lead to earning more money on the PGA TOUR, and these results have already been illustrated and confirmed using less refined data.

Meanwhile, there is a benefit to using less refined data in that it is easier to come by and available from more settings, thus allowing additional hypotheses to be tested. This chapter uses the more readily available year-end data from both the PGA TOUR and the Ladies Professional Golf Association (LPGA), in order to compare and contrast,

[1]This chapter draws from material in Shmanske, S. (2000). Gender, Skill, and Earnings in Professional Golf, *Journal of Sports Economics*, 1(4), pp. 385-400, Shmanske, S. (2004a) *Golfonomics*, (World Scientific Publishing Co., Inc., River Edge, NJ), and Shmanske, S. (2012b) *The Oxford Handbook of Sports Economics, Volume 2: Economics through Sports* eds. Shmanske, S. and Kahane, L. H., Chapter 3 "Gender and Discrimination in Professional Golf," (Oxford University Press, Inc., New York) pp. 39-54.

by gender, the relationships between skills and earnings. Keep in mind also that the annual data on individual performances by gender in PGA TOUR and LPGA golf are still head and shoulders above what is available in any other industrial setting. There is simply no other area where full, objectively-measured, productivity and earnings measures by gender are publicly available.

This chapter will use 2008 data from the PGA TOUR and the LPGA to update my previous research that used comparable 1998 data. In that earlier study, two methods of comparing men's and women's earnings showed that although women earned less, they were fairly (even more than fairly) paid once the skill levels evidenced in their performances were taken into account. By using the same methodology with newer data, the older results can be confirmed or contradicted and the extent to which gender discrimination in either direction changed over the period can be readily determined.

I begin with the recognition that if one group of people earns less than another, it may be because they are less productive, or because they are discriminated against, or both. In most industrial settings it is not possible to accurately measure all, or even any, of the relevant dimensions of individual productivity, therefore, it is impossible to determine the extent, or even the direction,[2] of any discrimination that may exist. The benefit of pursuing golf economics in this chapter is that the existence of high quality data on individual productivity allows one to untangle the productivity effect from the discrimination effect.

Shmanske [2000, 2004a] used the high quality data collected by the PGA TOUR and the LPGA during their 1998 seasons to examine the issue of gender discrimination in the earnings of professional golfers. Even though the men on the PGA TOUR earned more in total and more per tournament than the women on the LPGA, their extra earnings were justified because the men played more tournaments, played longer tournaments, and exhibited more skill. If anything, there was a bias in favor of the women who played in tournaments that were closed to men,

[2]Indeed, a person who earns 10% more than a colleague is actually discriminated against if the person produced 20% more than the colleague. Without knowing the underlying productivity it is impossible to determine which side (if either) is being discriminated against.

whereas the PGA TOUR events are open to all who can qualify regardless of gender. Indeed, a handful of women including Babe Zaharias, Annika Sorenstam, and Michelle Wie have competed in PGA TOUR events.

This chapter will update the analysis of gender discrimination in professional golf by using 2008 data to recreate the previous study. There may have been changes in the extent of discrimination in the past ten years, and by comparing recent data to those of 1998, the changes can be highlighted. For example, the National Committee on Pay Equity calculates a "wage gap," which is a raw measure of how much less women earn than men. According to their website, the gap narrowed by approximately 20% from 1999 to 2007, as women's earnings climbed from 72.2 cents to 77.8 cents for each dollar of men's earnings.[3] These calculations indicate that the wage gap is narrowing but still exists. But since these calculations do not control for productivity, education, or experience, they are actually a disservice to anyone truly attempting to understand the extent or direction of discrimination. These calculations are rhetorical demagoguery that may serve some political ends, but are useless with respect to scientific understanding. By using the sports data we can control for productivity and get a better picture of what has been happening to gender earnings differences over the past ten years.

The next section of the chapter explains the methodology used to examine the earnings gap. Then comes the description of the data, including summary comparisons between men and women for the years 1998 and 2008. The results follow and a brief summary concludes the chapter.

7.1 Methodology

This chapter will use regression analysis to measure the effect of each of the skills used by professional golfers to win prize money in tournaments. The specific skills and their measurements are described in the next section. The first set of regressions will use 2008 earnings per tournament as the

[3]http://www.pay-equity.org/info-time.html, accessed on 8/1/2009.

dependent variable. The regressions will be run for men and women separately:

$$Y_M = B_M X_M , \quad \text{and} \quad Y_W = B_W X_W ; \tag{7.1}$$

and pooled with an intercept-shifting dummy variable:

$$Y = BX . \tag{7.2}$$

In Eqs. (7.1) and (7.2), Y stands for income per tournament, X is a vector of relevant golf skills, and B is a vector of unknown coefficients interpretable as the price or value of each skill. In Eq. (7.1), the M and W subscripts refer to the split sample of men and women estimated separately. In Eq. (7.2) a dummy variable for women is added to the skill vector. These regressions will be compared to the same specifications from 1998. The sign and significance of the dummy variable will be the first indication of the extent of and change in gender discrimination.

It is well known in earnings regressions on levels that heteroskedasticity[4] usually exists. By transforming the dependent variable to natural logarithms,[5] this problem is avoided and a new (better) set of estimates can be obtained. Again, these regressions will be run on men and women separately and pooled, and the results compared to 1998.

The examination of discrimination based on the dummy variable in the above pooled regressions can be supplemented with a decomposition of the earnings gap following the method of Oaxaca [1973]. The equalities in Eq. (7.1) can be combined to yield:

$$Y_M - Y_W = (B_M - B_W) X_W + B_M (X_M - X_W) . \tag{7.3}$$

Since the regression lines go through the means of the data, Eq. (7.3) can be interpreted to mean that the difference between average earnings for

[4]This fancy word means that some individual observations are, predictably, measured with more error than others and should, therefore, be given less weight in the calculation of the regression coefficients.

[5]The transformation to logarithms scales back the larger values (the ones with predictably larger errors), and their associated errors, giving each observation, more or less, the same level of importance.

men and women can be broken down into two terms. The first term on the right hand side is the discrimination term. It is the difference in prices (B's)[6] paid to men and women, and weighted by the average skill level of the women. The second term is not discrimination, it is due to the differences in skills (X's) as weighted by the men's coefficients. The payoff to this earnings gap decomposition is that the total gap, $Y_M - Y_W$, is separated into its component parts, so much for each separate dimension of the skill vector, X, so much due to different implicit prices paid to men and women, and a general component associated with the constant term.

The Oaxaca decomposition is not without ambiguities. First, the weights in Eq. (7.3) could be reversed as Eq. (7.3) can be rewritten as:

$$Y_M - Y_W = (B_M - B_W)X_M + B_W(X_M - X_W). \tag{7.3a}$$

As Oaxaca notes, this is essentially an index number problem. Results for the decomposition in Eq. (7.3) are reported in Tables 7.6 and 7.7, but see the notes accompanying the tables for the alternate weighting in Eq. (7.3a).

Second, as Oaxaca and Ransom [1999] point out, the calculations are sensitive to the measurement of the independent variables in Eq. (7.1). The overall amount of discrimination is not affected, but its breakdown among the individual skill dimensions and the constant term is. Furthermore, the interpretation of the results is not straightforward when some of the skills have negative prices.[7] For this reason, each of the skill variables is transformed linearly, (for the purposes of the earnings decomposition, but not for the original regressions) so that zero measures the lowest level of

[6]The B's are referred to as prices in a metaphorical sense. Consider the driving distance variable which is one of the skills in the X vector. The golfer does not sell each yard of driving distance for a positive price in the usual sense of selling a product or even in the sense of working for a piece rate applied to each yard of driving distance. Rather, the extra driving distance statistically leads to higher earnings at a certain rate which is metaphorically referred to as a price. Economists will also use the jargon, marginal revenue product, to refer to the increase in earnings related to the increase in some input. The term, price, seems less awkward than marginal revenue product. Perhaps in standard English usage the term, payoff, would be more appropriate.

[7]For example, the skill of putting is measured as the number of putts taken on average on greens "hit in regulation" (explained below). Since fewer putts are better, the "price" of a putt is negative, that is, more putts translates into lower earnings.

each skill by anyone in the data, and all skills are measured positively. For example, suppose there were only two observations of the skill of putting consisting of the number of putts taken on greens reached in regulation. Golfer A took 1.72 putts per green and golfer B took 1.80 putts per green. In the transformed data golfer B is the worst and gets a 0. Golfer A gets a positive 0.08 because on average he takes 0.08 fewer putts per green. Thus, a higher measure indicates a higher skill and all prices will be positive. These linear transformations do not change the statistical properties of the regression equation or the size of the coefficients except for the constant term.

Ultimately, these ambiguities are not as troublesome as they might ordinarily be because the exact same methodology is used to assess the 1998 data as the 2008 data. Although no specific bias is expected, if some bias is present in both years, then its effect will cancel out and the trend in gender discrimination can still be discerned.

7.2 The Data

Professional golfers use a variety of skills to hit a small ball into a small hole in the ground that is sometimes over a quarter of a mile away. Although probably familiar to most readers, a short review of the skill measurement protocols and the names of the variables is included below.

A player's first stroke on each hole is called the drive and the object is to hit the ball as far as possible down a closely mowed area of the course called the fairway. There are two measures of this skill. DRIVDIST is the length in yards of the drive, and DRIVACC is the percentage of a player's drives that end up in the fairway. See Table 7.1 for the summary statistics for these variables for both years and both genders. The next shot is typically played to the green which is an even more closely mowed and smoothed area of grass where the hole is located. If the second shot on a par four hole is on the green, then the player is said to have reached the green in regulation. GIR is the percentage of times that the golfer achieves this, and thus measures the degree of this skill that the player exhibits. Once on the green in regulation, a player can take two putts to achieve par for the hole, but of course, the player is attempting to take only one putt.

Table 7.1 Summary statistics.

PGA TOUR 1998 top 130 money winners				
Variable	Mean	Standard Deviation	Minimum	Maximum
DRIVDIST	271.25	7.800	249	299.4
DRIVACC	69.77	4.931	52.4	80.4
GIR	65.62	2.679	55.5	71.3
PUTTPER	1.778	0.0223	1.724	1.865
SANDSAVE	52.46	5.749	36.2	69.8
EVENTS	26.15	4.357	15	35
Y1998 nominal	623,400	500,330	10,870	2,591,031
(constant $2008)	823,511			
YPEREVENT	25,123	22,065	374.7	112,700
(constant $2008)	33,187			
LNYPEREVENT	9.808	0.831	5.926	11.632

PGA TOUR 2008 top 130 money winners				
Variable	Mean	Standard Deviation	Minimum	Maximum
DRIVDIST	287.99	8.612	261.4	315
DRIVACC	63.41	5.154	51.1	74.0
GIR	64.91	2.633	58.0	71.1
PUTTPER	1.783	0.0227	1.718	1.842
SANDSAVE	49.92	5.628	36.8	63.7
EVENTS	26.26	4.464	15	36
Y2008	1,752,418	1,034,401	775,899	6,601,094
YPEREVENT	71,323	51,495	23,026	287,570
LNYPEREVENT	10.995	0.564	10.044	12.569

Table 7.1 Summary statistics (continued).

LPGA 1998 top 130 money winners

Variable	Mean	Standard Deviation	Minimum	Maximum
DRIVDIST	238.00	8.834	215.1	260.4
DRIVACC	70.00	5.697	56.3	83.2
GIR	65.79	3.862	55.6	78.1
PUTTPER*	1.815	0.0258	1.734	1.870
SANDSAVE	39.75	7.200	21.9	62.2
EVENTS	23.34	3.989	13	32
Y1998 nominal	186,010	195,630	27,720	1,092,748
(constant $2008)	245,719			
YPEREVENT	7,751	8,033	1,517	52,040
(constant $2008)	10,239			
LNYPEREVENT	8.576	0.835	7.325	10.860

LPGA 2008 top 130 money winners

Variable	Mean	Standard Deviation	Minimum	Maximum
DRIVDIST	248.08	8.605	224.8	269.3
DRIVACC	67.79	5.266	50.3	79.9
GIR	63.69	3.226	54.7	71.6
PUTTPER	1.820	0.0280	1.736	1.918
SANDSAVE	38.08	8.228	14.29	60.0
EVENTS	21.84	4.517	10	30
Y2008	398,871	439,304	57,015	2,763,193
YPEREVENT	19,299	21,672	2,441	110,528
LNYPEREVENT	9.351	1.013	7.800	11.613

Notes: The 1998 data come from Shmanske [2000]. The 2008 data come from the websites of the PGA TOUR and the LPGA. DRIVDIST is measured in yards. DRIVACC, GIR, and SANDSAVE are measured in percentages times 100. PUTTPER is measured in putts per green.

* In 1998 the LPGA did not keep the PUTTPER statistic. The values used are estimates of what PUTTPER would have been given the other statistics. See Shmanske [2000] for details.

PUTTPER measures the average number of putts taken by a golfer to finish a hole on which the green was reached in regulation. PUTTPER is negatively related to the skill of putting since the object is to take as few strokes as possible. If a player does not reach the green in regulation, the ball sometimes ends up in a sandy area known as a bunker and another special skill is required to extract the ball from the sandy area. This skill, called SANDSAVE, is the percentage of times that the player hits from a greenside bunker and sinks the next shot to finish the hole. These five skills are tracked by the PGA TOUR and the LPGA and form the basis for almost all of the statistical work on the relationship between skills and earnings in professional golf.

The 1998 data are taken directly from Shmanske [2000] where the description of the sources and calculations is documented. In this chapter we shall reprint the appropriate statistics from the tables therein. The 2008 data are taken from the websites of the PGA TOUR and the LPGA.[8]

There are four separate sets of data, two genders in each of two years. Each of the four sets has a different number of golfers for whom all of the statistics on skills and earnings were retrievable–for the PGA TOUR, 130 top golfers in 1998 and 197 in 2008, for the LPGA, 178 top golfers in 1998 and 159 in 2008. The regressions below use all of the available data. However, for the purposes of comparing the summary statistics, the clearest comparison comes from considering the same number of top golfers from each data set. Therefore, in each part of Table 7.1, only the top 130 money earners are included.

The table shows that the biggest changes occurring over the ten-year period are in the dollar amounts which jump dramatically for both the men and the women even after controlling for inflation. For example, the PGA TOUR earnings per event averaged $25,123 in 1998. To keep up with inflation this figure would have to increase to $33,187 in 2008. But the men did much better than simply keeping up with inflation as the average earnings per event increased to $71,323 by 2008. Men's average earnings, average earnings per event, and the earnings of the money leader all more than double in real terms. The women's earnings also increased, beating inflation by a wide margin, but did not quite keep up with the men's. The

[8]See http://www. pgatour.com and http://www.lpga.com.

leading money winner's earnings almost doubled as did the average earnings per event, but the average yearly earnings for the top 130 women increased by "only" 62% in real terms.

When it comes to assessing the skills changes over the past ten years, the obvious difference is a marked and statistically significant increase in driving distance of over 16 yards for the men on the PGA TOUR. This is coupled with a large, statistically significant decrease in driving accuracy. There are also decreases in the measures of the other skills, although the magnitudes of the decreases suggest that the changes are of lesser importance.

The changes on the LPGA mimic the men's results very closely. For driving, there is a large and statistically significant increase in driving distance (about 10 yards) and a decrease in driving accuracy. Additionally, there are decreases in the other three skills. The main difference between the men and the women is that the decrease in GIR for the women at just over two percentage points from 65.79% to 63.69% is of a magnitude that would seem to make an important difference.

Distance has improved for golfers at all levels of the game including top level professionals, senior professionals, serious amateur competitors, and casual weekend golfers. There is no question that major sources of the increased distance are the technology improvements in the design and materials used in golf balls and golf clubs. As a matter of pure geometry, the farther one hits a ball, the farther (in feet) it strays from its intended direction for any given positive deflection (in degrees) from its intended direction. Therefore, a natural outcome of hitting the ball farther is the decrease in driving accuracy noted in the statistics. Of course, since both distance and accuracy are desirable, golfers attempt to manage the tradeoff. There is room for more research on this one micro tradeoff involved in golf, but it is beyond the scope of this chapter.

Finally, in assessing the data to get a feel for the lay of the land it is interesting to compare the men to the women directly. The main thing that immediately pops out is that the men's driving distance is longer than the women's by a significant margin of 30 to 40 yards. And in distinction to the large difference in driving distance, the other skills measures are quite similar across genders. The women are slightly better in driving accuracy. Hitting greens in regulation is practically a dead heat. Perhaps surprisingly, the men are slightly better in putting, and even though the magnitude of the difference (1.78 to 1.82 for 2008) is small it adds up. For example, over a four-day 72-hole tournament with roughly two-thirds, or 48, of the greens reached in regulation the difference amounts to almost two strokes (0.04 times 48 equals 1.92).

7.3 Results

This section will compare results from 1998 to those of 2008. There are three sets of tests, taken in the following order. First, the section presents regressions of earnings per tournament measured in levels where the coefficients are interpreted as the prices of each particular skill. The test for discrimination is in the dummy variable in the combined regressions. Following this come the regressions of earnings per tournament measured in natural logarithms where the coefficients are interpreted as percentage changes in earnings due to each particular skill. Again, the test for discrimination focuses on the coefficient of FEMALE. Finally, a decomposition of the earnings gap based on the better-fitting logarithmic model is presented.

Consider Tables 7.2 and 7.3. Table 7.2 shows the 1998 regressions on levels for men and women separately, and combined with a dummy variable for women, as reprinted from my earlier research. To summarize, for the men, all the skills except SANDSAVE are significant in the theoretically predicted direction, and the estimates are of an economically meaningful magnitude. For the women, only GIR and PUTTPER are significant, and their values are less than those for the men. In the combined regression, DRIVDIST, DRIVACC, and PUTTPER are significant, and the dummy variable for the women indicates a bias **in favor** of the women of over $9,000 per tournament. That is, controlling for the lower skill levels exhibited in LPGA events, the LPGA pays $9,000 more per tournament than the PGA TOUR.

In 2008 there are some changes indicated in Table 7.3. For the men, all of the skills except DRIVACC are significant and of relevant magnitudes. In fact, the magnitudes of the significant coefficients all increase which is to be expected given the roughly 32% accumulated

Table 7.2 Regressions of earnings per tournament on skills, (t-statistics) 1998.

Equation	7.1	7.1	7.2
Sample	men only	women only	combined
Dependent Variable	YPEREVENT	YPEREVENT	YPEREVENT
Constant	334,476* (1.809)	117,193*** (4.000)	102,461 (1.516)
FEMALE			9,136.14* (1.777)
DRIVDIST	980.499*** (3.560)	47.1722 (0.724)	527.503*** (4.004)
DRIVACC	832.748* (1.729)	42.8813 (0.489)	501.202*** (2.611)
GIR	1381.27* (1.673)	656.092*** (5.069)	232.120 (0.812)
SANDSAVE	31.1046 (0.106)	35.3340 (0.663)	139.199 (1.179)
PUTTPER	-408,078*** (-4.704)	-92,432*** (-7.100)	-156,287*** (-5.246)
adjusted R^2	0.334	0.547	0.443
n	130	178	308

Notes: Source, Shmanske [2000]. *, **, *** indicate significance at the .10, .05, .01 levels respectively.

inflation over the period and the increase in the inflation-adjusted prize levels. SANDSAVE is now significant but accuracy with tee shots seems to be unimportant. The decrease in the player's accuracy with tee shots that was noted in the summary statistics is consistent with a rational response of golfers to the waning monetary importance of this skill.

Table 7.3 Regressions of earnings per tournament on skills, (t-statistics) 2008.

Equation	7.1	7.1	7.2
Sample	men only	women only	combined
Dependent Variable	YPEREVENT	YPEREVENT	YPEREVENT
Constant	490,196 (1.506)	88,856 (0.800)	304,015* (1.763)
FEMALE			9,257.93 (0.708)
DRIVDIST	1032.29* (1.704)	238.62 (1.011)	438.28 (1.277)
DRIVACC	-1000.20 (-1.045)	-230.27 (-0.655)	-838.01 (-1.571)
GIR	2792.49* (1.740)	1861.02*** (3.424)	1718.77** (2.044)
SANDSAVE	1994.33*** (3.093)	300.09* (1.836)	858.23*** (2.992)
PUTTPER	-531,893*** (-3.504)	-133,760** (-2.987)	-267,581*** (-3.731)
adjusted R^2	0.167	0.296	0.279
n	197	159	356

Note: *, **, *** indicate significance at the .10, .05, .01 levels respectively.

For the women, putting and hitting greens in regulation are still significantly related to earnings per tournament and SANDSAVE is added to the list of significant skills. Furthermore, the magnitudes of the coefficients increase markedly, almost tripling for the case of GIR.

In the combined regression, GIR, SANDSAVE, and, PUTTPER are significant and their coefficients increase by an average of over 400% from the 1998 figures. Interestingly, however, driving distance and accuracy are no longer important in the combined regression. Although the point estimate of about a $9,000 bias in favor of women remains the same as in 1998, the statistical significance is now gone. Therefore, this equation no longer rejects the hypothesis that the LPGA does not discriminate in favor of women in their tournament prize funds.

Overall, the 2008 regressions on levels of earnings per tournament are estimated with less precision than those from 1998. The levels regressions are easily interpretable, however, they suffer from heteroskedascity that is corrected by the logarithmic transformation we now move to.

Consider Tables 7.4 and 7.5. Table 7.4 recaps the 1998 results. In all three regressions, all of the coefficients have the expected sign and all are statistically significant except SANDSAVE for the men. An increase in a yard of driving distance can increase earnings by between 1% and 3%. An increase of a percentage point in DRIVACC or SANDSAVE can also increase earnings by a few percent. GIR is especially important in that an increase by one percentage point in this skill can increase earnings by 7% to 12%. Intuitively grasping the magnitude of the coefficient of PUTTPER involves a little computation on the side. The coefficient measures the effect of a decrease of one putt per green reached in regulation. Consider the effect of a decrease in one putt per tournament on the greens reached in regulation. Since tournaments typically have 72 holes with roughly 65% of them reached in regulation we must divide the coefficient of PUTTPER by 0.65 x 72 = 46.8 to get the effect. For the combined sample the effect of a decrease of one putt is about a 42% (that is, -19.6158 divided by 46.8 equals -0.419) increase in earnings per tournament. This figure is reasonable. For the leaders in a tournament, moving up one place in rank order can almost double one's payoff. Meanwhile, in the middle of the pack, a one-stroke improvement could increase one's rank from a ten-way tie for fortieth place into a ten-way tie for thirtieth place, with a commensurate increase in earnings. The bottom line for the 1998 equations is that the FEMALE dummy was insignificant, again failing to show discrimination against women in professional golf.

Table 7.4 Regressions of log earnings per tournament on skills, (t-statistics) 1998.

Equation	7.1	7.1	7.2
Sample	men only	women only	combined
Dependent Variable	LNYPEREVENT	LNYPEREVENT	LNYPEREVENT
Constant	23.8913*** (4.024)	31.0894*** (9.637)	28.6281*** (10.09)
FEMALE			0.2859 (1.324)
DRIVDIST	0.03624*** (4.099)	0.01574** (2.195)	0.02327*** (4.205)
DRIVACC	0.05112*** (3.307)	0.01683* (1.743)	0.02715*** (3.367)
GIR	0.07519*** (2.836)	0.12225*** (8.578)	0.11012*** (9.172)
SANDSAVE	0.00398 (0.422)	0.01585*** (2.700)	0.01203** (2.427)
PUTTPER	-18.3444*** (-6.588)	-19.9488*** (-13.92)	-19.6158*** (-15.68)
adjusted R^2	0.516	0.814	0.844
n	130	178	308

Notes: Source, Shmanske [2000]. *, **, *** indicate significance at the .10, .05, .01 levels respectively.

Now, moving up to 2008 and Table 7.5, we see that, as in 1998, the best fitting model is the combined equation with no evidence of a discrimination for or against women. The magnitudes of the coefficient estimates are very similar to those of 1998 for all of the variables except

Table 7.5 Regressions of log earnings per tournament on skills, (t-statistics) 2008.

Equation	7.1	7.1	7.2
Sample	men only	women only	combined
Dependent Variable	LNYPEREVENT	LNYPEREVENT	LNYPEREVENT
Constant	28.1582*** (4.819)	23.153*** (3.701)	26.6406*** (6.405)
FEMALE			0.1293 (0.410)
DRIVDIST	0.01671 (1.537)	0.0184 (1.387)	0.0169** (2.045)
DRIVACC	-0.00090 (-0.052)	-0.00107 (-0.054)	-0.0020 (-0.153)
GIR	0.07142** (2.479)	0.1100*** (3.595)	0.09473*** (4.670)
SANDSAVE	0.01987* (1.717)	0.01336 (1.451)	0.01549** (2.239)
PUTTPER	-15.6741*** (-5.753)	-14.286*** (-5.665)	-15.547*** (-8.988)
adjusted R^2	0.203	0.439	0.566
n	197	159	356

Note: *, **, *** indicate significance at the .10, .05, .01 levels respectively.

DRIVACC which is decidedly insignificant in all three equations. Each yard of driving distance can add about 1.7% to earnings. A percentage point increase in GIR leads to about a 9% increase. A percentage point increase in SANDSAVE adds between 1% and 2%. Finally, earnings are once again extremely sensitive to putting. Like 1998, there is no evidence showing discrimination against female professional golfers once skill levels are taken into account.

Based on the logarithmic models the unadjusted earnings gap at the mean of the data is 1.8216 log points in 1998 and 1.5485 in 2008. The change represents a decrease in the gap of 15%, a figure comparable to the 20% reduction in the gap calculated by the National Committee on Pay Equity. Nevertheless, this figure is unadjusted for skills levels, which have changed, and as such, may be misleadingly irrelevant. Indeed, a positive wage gap by itself can even be consistent with reverse discrimination as was the case in 1998. Using the method described in Eq. (7.3), two sets of terms can be identified to partition the total gap into portions attributable on the one hand to productivity and on the other to discrimination. The 1998 decomposition is listed in Table 7.6.

In the calculations underlying Table 7.6, the separate regressions in Table 7.4 are re-estimated with rescaled skills variables. Each of the skills variables is linearly transformed so that zero represents the worst level of the skill by any golfer and so that all skills are measured positively. This transformation does not effect the coefficients, except for a negative sign for PUTTPER and except for the constant term, or the statistical properties of the regression.

Table 7.6 confirms that women are not discriminated against. The first column represents the percentage of the earnings gap that is due to the estimated coefficients, which we have been calling prices, and, therefore, represents discrimination against women in the sense that the prices paid for their skills are lower than the prices paid to men. A positive number in this column represents discrimination against women. Meanwhile, a negative number in this column represents discrimination in favor of women. The totals at the bottom indicate that a **negative** 29% of the earnings gap is due to discrimination, meaning that based on the higher level of skills exhibited by the men, the earnings gap should be even 29% higher than it is. On a skill by skill basis, there is evidence that women's driving skills are rewarded less than men's by 32% for distance and 34% for accuracy. However, this deficit is more than made up by the higher payoff on the other skills (especially GIR) and by the 27% general favoring of women in the constant term.

Table 7.6 Decomposition of PGA TOUR - LPGA earnings differential 1998.

(Unadjusted differential = 1.8216)

Variable	% of Gap Due to Differences in Estimated Coefficients	% of Gap Due to Differences in Average Skill Levels
Constant	-0.272	0
DRIVDIST	0.320	0.689
DRIVACC	0.342	0.020
GIR	-0.463	0.066
SANDSAVE	-0.129	0.029
PUTTPER	-0.088	0.487
Total	-0.290	1.291

Notes: Source, Shmanske [2000]. The above table weights the percentage of the gap due to differences in estimated coefficients by the women's mean skills and the percentage of the gap due to differences in average skill levels by the men's estimated regression coefficients. When the alternative weighting scheme is used, the totals are -0.057 for discrimination and 1.055 for skills differences.

The second column in Table 7.6 indicates the percentage of the gap that is due to differences in the skill levels of men and women. So, for example, almost 69% of the earnings gap can be attributed to the longer driving distances exhibited on the PGA TOUR. As the total indicates, a full 129% of the earnings gap can be attributed to the greater skill levels implying that the earnings gap should even be 29% higher than it is.

As indicated in the note to Table 7.6, these figures may overstate the case. With the alternative weighting scheme from Eq. (7.3a), the discrimination **against men** is only on the order of five or six percent. As is usual for an index number of this sort, the truth is bracketed by the two extreme estimates.

Table 7.7 Decomposition of PGA TOUR - LPGA earnings differential 2008.

(Unadjusted differential = 1.5485)

Variable	% of Gap Due to Differences in Estimated Coefficients	% of Gap Due to Differences in Average Skill Levels
Constant	0.191	0
DRIVDIST	-0.025	0.452
DRIVACC*	0	0
GIR	-0.522	0.084
SANDSAVE	0.161	0.155
PUTTPER	0.103	0.401
Total	-0.092	1.092

Notes: The above table weights the percentage of the gap due to differences in estimated coefficients by the women's mean skills and the percentage of the gap due to differences in average skill levels by the men's estimated regression coefficients. When the alternative weighting scheme is used, the totals are -0.097 for discrimination and 1.099 for skills differences.
* DRIVACC is dropped from the regression. The coefficients were negative and the corresponding t-statistics were -0.052 and -0.054 for the men and women respectively.

Before moving to the discussion of the 2008 decomposition in Table 7.7, one change needs to be explained. In the 1998 comparison, all of the coefficient estimates are the right sign and all but one (SANDSAVE for men) are statistically significant. Therefore, some care should be exercised when interpreting the SANDSAVE portion of the decomposition. Upon examination, SANDSAVE does not seem to be wildly influencing the results. In the 2008 equations, however, the point estimate of the return to DRIVACC is actually negative, although statistically insignificant, and, quantitatively, practically zero. Since negative prices confuse the interpretation of the earnings gap decomposition, and since the coefficients are practically zero anyway, the regressions in Table 7.5 are re-estimated, with rescaled variables as explained above, and excluding DRIVACC. The coefficient estimates and the average skill levels are then used as in Eq. (7.3) to decompose the gap, and these measures are normalized by the average gap of 1.5485 log points to derive the percentage estimates in Table 7.7.

The decomposition in Table 7.7 confirms and reinforces the 1998 results. Women are not discriminated against in professional golf. There are minor amounts of discrimination in the returns to putting and SANDSAVE, on the order of 10% and 16% respectively. But these are more than offset in that the women are actually compensated a little more for the skill of driving distance and a great deal more (52%) for hitting greens in regulation. Meanwhile about 45% and 40% respectively of the earnings gap is explained by the greater skill levels of men in the areas of driving distance and putting. Overall, only **negative** nine percent of the gap between the women's average earnings and the men's average earnings is attributable to discrimination. That is, once the higher level of skills of the men is taken into account, the earnings gap should actually be nine percent higher.

7.4 Summary

High quality data on the skills required to excel in professional golf allow one to examine gender discrimination in society in a way that controls for productivity. Such is one of the important payoffs to studying economic issues through the lens of sports. Wage or salary gaps between men and women still exist in American society, although they have been narrowing. Nevertheless, such gaps are not evidence of discrimination in and of themselves. The gaps are justified if they are due to productivity differences.

In professional golf, regressions of earnings per tournament on the important skills, as identified by golf production functions, had indicated that there was no discrimination against female golfers in 1998. This chapter updates the analysis to 2008 data and reaches the same conclusion.

If anything, point estimates indicate that even a larger gap could be justified once the skills differences are taken into account. This result is reached by including a dummy variable in regressions that pool PGA TOUR and LPGA data and by performing a decomposition of the earnings gap by manipulating the coefficients from regressions that segregate the data. The pooled regressions indicate that women earn about $9000 or 13% too much per tournament. The decomposition indicates that the wage gap could be about nine percent higher than it is. However, none of these point estimates are precise enough to reject the null hypothesis that there is no discrimination either way.

The result of no discrimination should not be surprising. The PGA TOUR, which is the source of the data for the male golfers in the sample, does not discriminate against women. Women are allowed to play in PGA TOUR events and occasionally do so. Women are also allowed to attempt to qualify for PGA TOUR events based on their performance in qualifying tournaments. However, the competition is stiffer on the PGA TOUR than it is in the female-only LPGA tournaments. Therefore, female professional golfers find it more lucrative to compete in LPGA events where men are excluded. If the LPGA did not offer at least as good a deal to women as the PGA TOUR offers (in fact the LPGA offers a slightly better deal), then female golfers would leave the LPGA and attempt more often to compete on the open to all genders PGA Tour.

Chapter 8

Gender and Driving Distance[1]

The previous chapter showed how skills lead to earnings in professional golf, how the returns differ by gender, and what has changed over the ten-year period from 1998 to 2008. Remarkably, men and women are very close in the measurement of many of the skills. Perhaps surprisingly however, men are significantly better putters. The margin is small but it adds up because putting is such an important part of the sport. And men's relative superiority in putting is a finding that is robust over the ten-year period. Although there may be a subconscious mechanical reason,[2] I suspect that men are better because they practice more because there is more at stake in terms of prizes in the tournaments that they typically play. Remember, of course, that women could play for the same stakes as men but instead choose to play only against other women in LPGA tournaments where the prizes are lower. This is not an irrational choice by women, rather, it is the realistic choice if women cannot keep up or catch up to men in all of the skills, especially the skill of driving distance.

In a sport like golf, brute strength is not as important as it is in many other sports. Unlike football, weightlifting, or many other sports where

[1]This chapter draws from Shmanske, S. (2013) *Handbook on the Economics of Women's Sports* eds. Leeds, E. M. and Leeds, M. A., Chapter 4 "Gender and Skill Convergence in Professional Golf," (Edward Elgar Publishing Ltd, Northampton, Massachusetts) pp. 73-91.

[2]It has been suggested to me (but I cannot remember from where) that males have a better natural feel for putting because they have more experience with the parabolic falloff of a trajectory, and how it varies with the initial force–experience gained from observing their urination patterns. Seriously!

physical size and brute upper body strength are important factors to success, golf would seem to be a sport in which practice, repetition, hand-eye coordination, and timing would be of utmost importance. Arguably, women should be able to compete equally with men in these areas. Consequently, all else equal, women might be expected to compete effectively with men on an equal basis. Occasionally, (Babe Zaharias, Annika Sorenstam, Michelle Wie), women have competed directly against men but these are the exceptions. The vast majority of times women compete separately in women's only events. It is quite possible that this is an artifact of antique social mores that are slowly changing. As women (and society in general) become accustomed to the idea of women competing directly with men, women will be more willing to devote the practice time necessary to develop the level of skills to compete in gender neutral tournaments at the highest level. Many golf fans, including this author, are anxiously awaiting this day. In the meantime, however, as shown in Shmanske [2000, 2004a, 2012b, 2013], and the previous chapter, at their current skill levels women can earn more money by competing in women's only tournaments sponsored by the LPGA than they can by competing in the open to both genders tournaments on the PGA TOUR.

This chapter will compare men's and women's skill levels over the most recent 20-year period focusing on the skill of driving distance. Even though the skills of chipping, putting, and overall accuracy are also necessary, the skill of driving distance is especially interesting to look at for a number of reasons.

First, if women can catch up to men in driving distance, the skill most likely to be affected by overall size and brute upper body strength, they can surely, with practice, match men on the other skills. These other skills seemingly require touch, hand-eye coordination, flexibility, balance, and timing, all developed by practice, and all theoretically gender neutral. Women's performance relative to men in driving distance should be a prime determinant of women's ability to compete with men.

Second, the skill of driving distance has co-evolved rapidly with the growth of technology in golf equipment. Advances in materials and design for both golf balls and golf clubs have led to an increase in driving distance for golfers of all ages and abilities. If these new materials and designs differentially advantage women over men or men over women, it may advance or forestall, perhaps indefinitely, the day when women achieve parity with men.

Third, it is most obvious to even casual observers of the sport that the skill of driving distance truly separates professional golfers (men and women) from the rest of us. Even mediocre amateurs will occasionally have a great round, sink a long putt, chip in from off the green, or make a birdie, but most will never hit a drive over 300 yards long. Even though more astute fans and more accomplished amateurs recognize that professionals are perhaps even farther ahead in the areas of spin rate, trajectory control, and consistency, than they are in pure distance, these more subtle aspects of the game lack the "wow" factor that long drives engender. After all, the old adage "drive for show, putt for dough," comes from somewhere. And because of the fans' appreciation of the prodigious distances that professionals can hit the ball, there are important linkages from driving distance to prize funds to professional golfer earnings and back to the incentive to increase driving distance. These linkages might differ systematically between the PGA TOUR and the LPGA.

The remainder of this chapter is divided into four sections. The next section sets the stage for this research by presenting the previous estimates in the literature of the effect of driving distance on earnings in professional golf. The two following sections describe and carry out two separate analyses of the gender differences in driving distance. The first of these looks at whether women are catching up to men in the skill of driving distance. The second performs a Granger [1969] causality test to determine the extent to which increasing prize funds bring forth more driving distance, and the extent to which increases in driving distance provide fan enjoyment ultimately leading to larger purses. The last section summarizes.

8.1 Driving for Dough

Chapter 6 covered the professional golfer's earnings function in depth, looking at each of the five golf skills that form the basis for all of the statistical analysis of the golf production function, namely, driving distance, driving accuracy, greens in regulation, sand saves, and putts per green in regulation. Here, the focus is specifically on the numerous estimates of the effect of driving distance. Many studies have been undertaken, differing slightly based on the year and tour of the sample,[3] the format of the dependent variable, the functional form, and the list of included skill and control variables on the right hand side.

Some studies [Davidson and Templin, 1986; Nix and Koslow, 1991; and Shaffer et. al., 2000] came from the exercise, sport, and body mechanics tradition as opposed to the economics of sports field. While these studies hint at the importance of driving distance they do not report useable estimates of the effect of driving distance on scoring or on earnings. The earliest study that reports useable estimates [Shmanske, 1992] comes from the economics tradition. In that paper I used data on 60 top money earners from the 1986 season and regressed levels and natural logarithms of earnings per year and earnings per tournament on the five skills mentioned above, plus a calculated measure of short game skill and a measure of experience. In the levels specification the coefficient estimates are interpretable as values of the marginal products (VMPs) of the various skills. Thus, an extra yard of driving distance is worth an extra $6775 per year or an extra $341.10 per tournament according to the point estimates. Is this a little, or a lot?

To best answer the preceding question economists like to calculate what is known as an elasticity, which is simply a ratio of percentage changes. For example, one extra yard of driving distance is only a small percentage of the total distance. If this were to lead to a large percentage change in earnings, then it would be understandable that heroic efforts would be undertaken to increase driving distance, even if only by a small percentage. Alternatively, if a large increase in driving distance led only to a small increase in earnings, then the skill would not be that important. So for the skill of driving distance economists calculate the elasticity of earnings with respect to driving distance as the percentage change in

[3]The LPGA is open to women only. The Champions Tour, formerly the Senior PGA TOUR is open to anyone over the age of 50. The PGA TOUR is open to all comers but generally only men under the age of 50 compete because the exclusive nature of the other tours provides a better return per unit of skill due to the relatively lower skills of older men and women who are the other competitors. There are also European and other foreign-based tours and "minor" tours. These have been studied also, but the data availability has been best for the three major professional North American tours.

earnings divided by the percentage change in driving distance. The average drive in the 1986 sample in Shmanske [1992] is 262.1 yards, so a one-yard increase is actually an increase of approximately 0.38 percent (1/262.1 times 100 equals 0.38). Meanwhile, average yearly earnings in the 1986 sample were $256,000, so the increase of $6775 per year represents an increase of approximately 2.6 percent (6775/256,000 times 100 equals 2.64). Thus, the ratio of these percentage changes is the elasticity which in this case is approximately 6.94 (2.64/0.38). Thus, the data suggest that a one percent increase in driving distance would lead to an almost seven percent increase in earnings. See Table 8.1 for a listing of the various parameter estimates for driving distance that have been achieved in the economics literature, as well as a transformation of these estimates into elasticities for ease of comparison.

Another early study is Sommers [1994], who uses the full set of 183 PGA TOUR golfers for 1992. Sommers employs a slightly different arithmetic by using the natural logarithm of earnings per tournament as the dependent variable. Intuitively, changes in a natural logarithm can be interpreted as percentage changes, so that the parameter estimates of the right hand side variables are already in percentage change form. He discovers with high precision that one yard of driving distance is worth 0.038 log points, which means that one yard of extra distance on average will increase earnings by 3.8 percent. To turn this into an elasticity we must first determine how one yard of distance translates into a percentage change. For example, in this case, the average driving distance was 260.4 yards, so one yard is actually an increase in driving distance of 1/260.4 = 0.00384 or 0.384 percent. The corresponding ratio of these percentage changes (3.8/0.384 = 9.90) yields an elasticity of earnings per tournament with respect to driving distance of 9.90.

Several other studies examine variations on the theme of estimating VMPs or arrive at similar regression equations independently. These studies examine different samples or include different control variables in addition to the usual five skills of driving distance, driving accuracy, greens in regulation, sand saves, and putts per green. Moy and Liaw [1998] develop separate estimates of the parameters for the PGA TOUR, the Senior PGA, ane the LPGA for the 1993 season. Driving distance is statistically significant on the PGA TOUR and the Senior PGA but not in

Table 8.1 Estimates of driving distance and earnings.

Author and date	Sample	Descriptive Statistics[a]	Result	Elasticity[b]
Nix and Koslow [1991]	PGA TOUR 1987	$267,000 / year 263.4 yards	Estimates not reported	
Shmanske [1992]	PGA TOUR 1986	$256,000 / year 262.1 yards	1yd. → $6775**/year	6.94
Sommers [1994]	PGA TOUR 1992	260.4 yards	1 yd. → 0.038***ln$/event	9.90
Wiseman et.al. [1994]	PGA TOUR 1992 Senior Tour[d] 1992 LPGA 1992	70.9 strokes/round 261 yards	1yd. → -0.18*strokes/rd.	-0.66[c]
Moy and Liaw [1998]	PGA TOUR 1993	$265,000 / year 260.2 yards	Reports elasticities	9.28***
	Senior Tour[d] 1993	$329,000 / year 254.1 yards		3.14**
	LPGA 1993	$ 91,000 / year 226.9 yards		4.07
Berry [1999]	PGA TOUR 1999	71.84 strokes/round 272.2 yards	1yd. → -.175strokes/rd.	-0.66[c]
Shaffer, et.al. [2000]	PGA TOUR 1998 Senior Tour[d] 1998 LPGA 1998		Estimates not reported	
Shmanske [2000]	PGA TOUR and	$623,000 / year 271.25 yards	1yd. → .036***ln$/event	9.76
	LPGA 1998	$139,440 / year 236.6 yards	1yd. → .016ln$/event	3.79
Rishe [2001]	PGA TOUR and	$923,000 / year 273.2 yards	1yd. → .015ln$/event	4.10
	Senior Tour[d] 1999	$567,000 / year 265.9 yards	1yd. → 0ln$/event	0
Nero [2001]	PGA TOUR 1996	$433,000 / year 266.9 yards	1yd. → .05***ln$/year	13.34

Table 8.1 Estimates of driving distance and earnings (continued).

Author and date	Sample	Descriptive Statistics[a]	Result	Elasticity[b]
Fried et. al. [2004]	PGA TOUR 1998	$18,806 / event 270 yards	1yd. → $1,583 / event	22.73
	Senior Tour[d] 1998	$17,749 / event 262 yards	1yd. → $ 257 / event	3.79
	LPGA 1998	$ 5,598 / event 237 yards	1yd. → $ 434 / event	18.37
Alexander and Kern [2005]	PGA TOUR 1992-2001	$306,600 / year[e] 268.19 yards	1yd. → $11,728***/year 1yd. → 0.04*** ln$/year	10.26 10.73
Pfitzner and Rishel [2005]	LPGA 2004	$ 10,370 / event 249.8 yards	Reports elasticities	4.95***
Callan and Thomas [2007]	PGA TOUR 2002	$953,000 / year 280 yards	Reduced form: 1yd. → $12,257/year	3.60
			Multi-equation: 1yd → $5,892/year	1.73
Shmanske [2008]	PGA TOUR 2006	$71,258 / event 289.1 yards	Reduced form: 1yd. → .0326***ln$/event	9.42
			Structural model: 1yd. → .016 ln$/event	4.63
Kahane [2010]	PGA TOUR 2004-2007	$52,020 / event[f] 288.6 yards	OLS: 1yd. → $926**/event Median 1yd. → $407*/event 90th% 1yd. → $1,602*/event	5.14 2.26 8.89
Shmanske [2012b]	PGA TOUR and	$71,323 / event 288.0 yards	1 yd. → $1032*/event	4.17
	LPGA 2008	$19,299 / event 248.1 yards	1 yd. → $238.6/event	3.07

Notes: *, **, *** denote statistical significance of estimate at .10, .05, and .01 levels where reported in original paper.

a. Sample averages for earnings are measured in current dollars unless otherwise noted.
b. Implied elasticity of earnings (per year or per event) with respect to driving distance unless otherwise noted.
c. Elasticity of strokes per round with respect to driving distance.
d. The Senior Tour is now called the Champions Tour
e. Constant dollars 1982-4.
f. Constant dollars 2007.

LPGA sponsored events. Nero [2001] uses the highest 130 earners from 1996 to tightly resolve a parameter estimate that implies a high elasticity of earnings per year with respect to driving distance of 13.34. Additionally, Alexander and Kern [2005] use 10 years of PGA TOUR data in an unbalanced panel to track changes in the VMP's of golfer skills over the period from 1992-2001. Interestingly, the return to driving distance roughly tripled over this period almost exactly matching an approximate tripling in purse sizes.

The literature having demonstrated the robustness of the single equation approach of regressing earnings on skills, Shmanske [2000] pushed the parameter estimates by comparing the PGA TOUR to the LPGA for the 1998 season using the decomposition method of Oaxaca [1973]. In variations on this approach, Rishe [2001] decomposed the age-based earnings gap between the Senior PGA and the regular PGA TOUR using data from 1999, and Shmanske [2012b] re-examined with 2008 data the gender-based earnings decomposition of his earlier study as recapped in the previous chapter.

Others started to estimate multiple equation structural models. In these models golf skills do not directly produce earnings, rather, they produce scores in competitions which produce earnings according to a rank-based distribution of prizes. Early on, Wiseman *et al.* [1994] had estimated the first half of the structure with 1992 data. They found that driving distance was significant only for the PGA TOUR. One yard of distance led to a decrease in strokes per round of 0.18. This translates to an elasticity of strokes per round with respect to driving distance of -0.66, but since they do not consider the effect of this on earnings, the elasticity of earnings with respect to driving distance cannot be calculated. With 2002 data, Callan and Thomas [2007] extended this approach by estimating a three-step structural model in which skills produced scores, scores produced ranks in tournaments, and ranks in tournaments produced earnings. Combining the results of three equations, their estimate of the VMP of driving distance is that one yard increases yearly earnings by $5,892. This is less than half of their reduced form estimate of the same VMP, $12,257 per year, which they calculate for comparison sake.

Each of the above papers has used the yearly averages of earnings and skills as the basic units of analysis. Other work has attempted to improve upon the measurement of the skills or the calculation of the coefficients. Berry [1999] recognized that the skills may be interdependent in his study of PGA TOUR data from the first 28 tournaments of 1999. For example, the measure of sand saves depends upon how good a putter one is, the measure of greens in regulation depends on how long and accurate one's drives are, and, because of interaction, the value of driving distance depends on driving accuracy. Berry constructed new measures of each skill and included the interaction term between driving distance and driving accuracy in his regression. Thus, one yard of driving distance reduces strokes per round by 0.067 directly and another 0.108 at the mean level of driving accuracy.

As reported on in Chapter 6, Shmanske [2008] followed the top 100 PGA TOUR earners from 2005 during the 2006 season to gather data on a tournament-by-tournament basis. Two improvements follow. First, the actual measurements of the skills are adjusted for course characteristics. Second, instead of only having the year-end average skill level, a distribution of skills (from which mean, variance and skewness are calculated) is captured for each golfer. Then, following Scully's [2002] two-step suggestion, the distribution of skills produces a distribution of scores which, in turn, produces earnings per tournament. In the reduced form estimation, one yard of driving distance increases the natural logarithm of earnings per tournament by 0.0326 log points, whereas in the complex structural model one yard of driving distance affects mean, variance, and skewness of the scoring distribution for a combined effect on earnings of 0.016 log points, less than half of the reduced form estimate.

Kahane [2010] further improved our understanding of the golf production function and the value of driving distance by estimating a quantile regression. He used data from the PGA TOUR from 2004-2007. At the mean of the data an O.L.S. regression estimates that one yard leads to extra earnings of $926 per tournament. In the quantile regression at the median of the data one yard leads to $407 per tournament, but at the 90th percentile one yard leads to $1,602 per tournament of extra earnings.

Several studies [Wiseman *et al.*, 1994; Moy and Liaw, 1998; Shaffer *et al.*, 2000; Shmanske, 2000, 2012; Fried *et al.*, 2004; and Pfitzner and Rishel, 2005] have focused specifically on or at least included the LPGA in the research. These papers largely follow the same methodology of regressing some form of earnings on a set of skills and other control variables. Not surprisingly, the reported results mirror those for men. Although driving distances and average earnings are lower, there is a growth in both driving distance and earnings through time and a positive elasticity of earnings with respect to driving distance. For women, the elasticities range from 3.07 to 18.37.

To summarize the results of these studies with respect to driving distance as listed in Table 8.1, simply note that with one exception driving distance always has an elastic effect on earnings. That is, for any given percentage increase in driving distance, the effect on earnings is to increase them by an even larger percentage. Except for the one outlier in which driving distance has a zero effect on earnings, the elasticity estimates range from 1.73 to 22.73. For these 24 calculations of the elasticity of earnings with respect to driving distance the overall average is 7.42. Therefore, it would be a good guess to say that a one percent increase in driving distance would lead to more than seven percent increase in earnings.

8.2 Recent Trends in Driving Distance

A casual look at Table 8.1 confirms that the average driving distance of professional golfers has increased over the period in which economists have been studying the golf production function. A deeper look at the papers themselves indicates that there has not been a marked increase in the other skills over the same period. This is true for both genders, although the LPGA has been studied less and the evidence is spottier. This section takes a close look at the actual trends in driving distance for men and women over the last two decades, attempting to ascertain whether women are catching up or falling further behind in this skill.

Data was collected on the age and yearly average driving distance of male and female professional golfers on the PGA TOUR, the LPGA, and the Champions Tour from 1992 to 2010. In the data, 630 different women and 694 different men are represented. The following equation was estimated for women:

$$\text{DRIVDIST} = b_0 + b_1\text{AGE} + b_2\text{AGE}^2 + b_3 1993 + \cdots + b_{20} 2010 + e\,, \quad (8.1)$$

where i and t subscripts indicating the golfer-year of the individual observation are suppressed. The dependent variable, DRIVDIST, is the yearly average distance in yards of the golfer's tee shots. AGE and AGE^2 are included to capture a possible inverted-U shape age profile of the skill. An inverted-U shape means that driving distance may be increasing at early ages, but that there will be an exaggerated fall off of driving distance at the most advanced ages. A diagrammatic exposition will show a curve that is first sloping upwards and then starts to fall off. To get this pattern the coefficient of AGE will be positive, giving the upward sloping part, and the coefficient of AGE^2 will be negative to capture diminishing distance at older ages. The remaining variables are the years which are dummy variables that capture the recent improvement in the skill without imposing a specific functional form. The dummy variable for 1992 is omitted so that the b_3 through b_{20} coefficients measure the improvement over the base year of 1992.

A variation of this equation was estimated for men with the addition of a dummy variable, CHAMP, for golfers on the Champions Tour. The equation is also estimated on the pooled (both genders) data with the addition of the dummy variable, CHAMP, and a new dummy variable PGATOUR, to separate the men's distance on the regular and senior tours from the women's distance on the LPGA. The results appear in Table 8.2.

A quick look at the constant terms indicates that men out-drive women over this period by about 40 yards. This result is corroborated by the dummy variables in the pooled regression which indicate a 37 or 38 yard advantage for the men. With respect to the men themselves, controlling for age, there is only a one or two yard difference between the regular PGA TOUR and the Champions Tour. This result is confirmed by the closeness of the two male dummies in the pooled regression and the small size of the CHAMP variable in the male regression.

A quick look at the AGE and AGE^2 variables shows an increasing falloff of driving distance as age increases, essentially the latter half of the inverted-U shape age profile, although there is no strong evidence of the upward-sloped part of the inverted-U. The result is consistent across gender. This result is not surprising in light of the fact that the upward sloping part of the inverted-U age profile probably occurs for golfers at ages younger than any that are captured in the data set. As such, a level

Table 8.2 Coefficient estimates (t-statistics) of Eq. (8.1).

Dependent Variable: DRIVDIST

Sample:	Women	Men	Pooled
Constant	229.647(84.06)***	269.803(105.05)***	227.118(120.35)***
AGE	0.021(0.14)	-0.102(-0.73)	0.129(1.25)
AGE^2	-0.006(-2.86)***	-0.005(-2.63)**	-0.007(-5.12)***
CHAMP		-1.635(-2.07)**	37.258(54.38)***
PGATOUR			38.138(173.1)***
1993	1.706(1.66)*	-0.002(-0.00)	0.932(1.39)
1994	2.677(2.60)***	1.868(2.25)**	2.339(3.47)***
1995	7.842(7.71)***	3.524(4.30)***	5.902(8.87)***
1996	9.354(9.22)***	6.649(8.16)***	8.120(12.24)***
1997	12.992(12.03)***	8.183(10.13)***	10.318(15.14)***
1998	13.502(12.54)***	11.215(13.81)***	12.171(17.83)***
1999	12.679(11.00)***	12.929(16.08)***	12.612(18.07)***
2000	14.800(13.76)***	14.029(17.58)***	14.225(21.05)***
2001	18.060(16.77)***	20.003(25.11)***	18.946(28.05)***
2002	23.476(21.83)***	20.232(25.79)***	21.434(32.05)***
2003	24.689(24.39)***	26.621(33.66)***	25.525(39.07)***
2004	23.766(23.48)***	27.189(34.53)***	25.334(38.88)***
2005	21.643(21.20)***	29.068(37.31)***	25.321(38.91)***
2006	23.016(22.55)***	29.333(37.46)***	26.092(40.02)***
2007	23.151(22.44)***	28.930(37.18)***	26.038(39.86)***
2008	21.692(21.03)***	27.497(35.47)***	24.585(37.73)***
2009	24.352(22.94)***	28.053(36.09)***	26.176(39.61)***
2010	21.660(20.32)***	28.127(36.55)***	25.095(38.15)***
adj. R^2	0.428	0.641	0.835
n	3823	4115	7938

Note: *, **, *** denote statistical significance of estimate at .10, .05, and .01 levels.

part of maximized driving distance followed by a fall off at advanced ages is exactly what occurs with an insignificant coefficient of AGE and a significant negative coefficient of AGE^2.

As far as the increased distance over time goes, both men and women have gained considerable distance over the period. Women have gained over 20 yards on average while men have gained closer to 30 yards. Interestingly, however, the pattern and timing of the gains differs by gender. Early in the period women gained more than men closing the distance gap by almost 5 yards by 1997. However, from 1997 until 1999, women's distances leveled off while men continued to gain, both ending up about 12 yards longer than in 1992. Both genders made large gains in the three years following 1999 with women adding another 11 yards for a total gain of over 23 yards compared to 1992, while men added 8 yards for a total gain of 20 yards. So from 1992 to 2002, women gained more yards on their drives than men did, making a small inroad on closing the gender gap in driving distance. Women were over 23 yards longer in 2002 than in 1992, while over the same period the men gained only a fraction over 20 yards. Men, of course, started from a higher base in 1992, but if the skill levels were converging this is precisely the pattern of coefficients one would expect.

However, the skill convergence disappeared after 2002 as the women's driving distance more or less leveled off while the men's driving distance made over a six-yard leap from 2002 to 2003 before leveling off. By 2010, men were 28 yards longer than in 1992 while women had gained less than 22 yards. The comparative pattern of gains to driving distance over the period is shown graphically in Fig. 8.1. The first half of the top panel might be construed to indicate that women's driving distance was converging upwards toward the men's driving distance, but the second half of the top panel shows that the gains were lost. The bottom panel puts into perspective the relatively unchanged gender driving distance gap over the period.

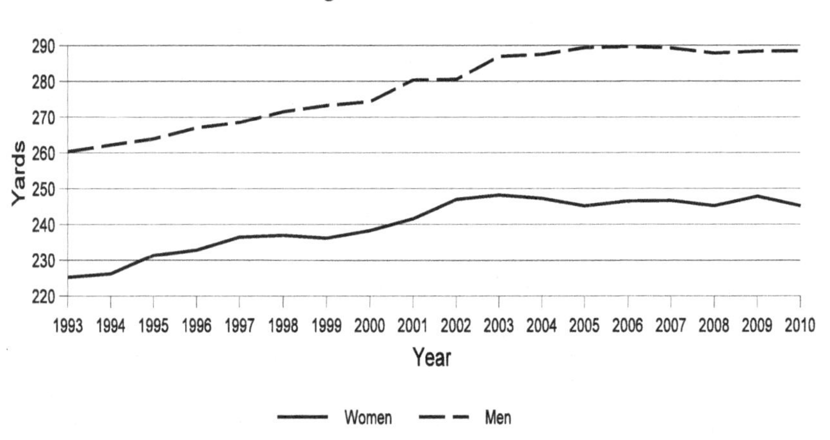

Fig. 8.1 Driving distance by gender.

8.3 Purses, Distance, and Granger Causality

It is obvious that both driving distance and prize funds have increased dramatically over the past two decades. Hypothetically, there can be causality going in both directions between these two time series, as well as exogenous influences on both variables. For example, the growth in prize funds means that there is a greater incentive to practice all golf skills, thus supplying the underlying theoretical argument for why increases in prize funds cause increases in driving distance. At the same time, however, increases in driving distance have the ability to attract more fans thus allowing tournament promoters to offer higher purses. This is why increasing driving distance may increase the prize funds. Alternatively and/or simultaneously, an outside factor may be causing both trends. For example, it may be that the arrival on tour of popular and charismatic golfers, such as Tiger Woods and John Daly, who also happen to be long drivers, has simultaneously led to increases in purses and average driving distances in professional golf.

It is beyond the scope of this chapter, and possibly beyond the limitations of the available data, to sort out all the possible directions of causality in a fully-specified simultaneous equation model. It is possible, however, to get a sense of what is going on from a statistical look at the timing of the growth in the related time series, in what is called a Granger causality test. The simple intuition is that if long drives, perhaps from technological innovations, were creating the fan interest and leading to higher prize funds, then the long drives should appear before the increases in the prize funds. If the increased prize funds were creating the incentive to practice more to develop higher skills, then the prize funds should go up first.[4] In reality, both arguments may be operative so that discovering the strength of each cause and effect becomes muddled. Furthermore, everyone is cognizant of avoiding the *post hoc ergo prompter hoc* fallacy so that claims of actual causality do not come from a simple comparison of the timing of effects in the time series data. What is discoverable from this type of analysis is called Granger causality, the procedure to be explained more below. Granger causality can never prove causality for the same reasons that statistical correlation can never prove actual causation. The only bottom line result of a Granger causality test is that evidence consistent with true causality is either found in the data or not.

[4]If the increased prize funds are anticipated, then the driving distance skill could be developed beforehand thus leading to contemporaneous increases, but there would seem to be little reason to increase the skill *because of the money* even before the money arrives.

Notwithstanding the theoretical problems of interpreting Granger causality, it will be interesting to compare the results obtained for the LPGA to those for the PGA TOUR. On the one hand it is possible that the wow factor of long drives leads to higher purses on the PGA TOUR, where in fact, the drives are the longest, but not in LPGA tournaments. On the other hand, women professionals still impressively outdrive casual amateur golfers creating considerable entertainment value for the fans of the sport which could lead to increases in prize funds for the LPGA. There also may be a gender-based difference in the monetary returns to developing driving distance. One might argue that the higher prize funds on the open to all genders PGA TOUR supplies the same incentive to both men and women to develop longer tee shots. However, it is the marginal return to increases in driving distance that is important, and this could be even higher (for the women) in LPGA events. Even though the evidence captured in Table 8.1 shows that the return to driving distance is higher on the PGA TOUR in both 1998 and 2008, the effective return, conditioned on women's average skill levels, could be higher on the LPGA than on the PGA TOUR where it might approach zero. The bottom line is that any pattern of Granger causality consistent evidence could be uncovered, making it important to look at what the data says.

Essentially, the Granger test has two parts. First, the first differences in each data series are calculated by subtracting the prior year's amount from the current year amount. For example, if driving distance in year t is 270 yards and in year t-1 is 268 yards, then the first difference is two yards. Then, the first differences are regressed on their own lagged values to determine what the statistical lag structure looks like. Second, keeping the significant own lags (call it time series "A") on the right hand side, the lags of the other time series (call it time series "B") are added to see if they can add any significant explanatory power. If they do, then series A is said to be Granger-caused by series B. The intuitive idea is that something happening to series B has a significant effect that shows up later on series A. If none of the lagged differences of B are significant, then there is no evidence that B Granger-causes A. This would be the case if there were no causality flowing from series B to series A. The Granger causality tests were performed separately for men and women on the time series of average driving distances and yearly total purses. The results are in Table 8.3.

The columns in the bottom panel of Table 8.3 report the results of four regressions. They will be taken in turn. The first column examines whether purses can be claimed to Granger-cause distance for men. First, in regressions not reported, it was determined that for the change in driving distance for men, the second own lag was significant but not the first. This means that something, perhaps a change in technology, may happen to improve driving distance in one year, and that it will also have a significant effect two years down the road. For example, perhaps a technological innovation occurred in one year and started to diffuse only slowly through the ranks of professional golfers in the next year but was fully diffused through the ranks by the second year. After discovering the pattern of driving distance changes the addition of lagged changes in purses are added to see if they are significant. In this case no additional explanatory power is added. Thus, it is shown that the growth in purses does not significantly increase driving distance in the Granger sense.

But now consider the second column which tests whether driving distance Granger-causes purse growth for men. In the auxiliary regression performed first, it was determined that only the first lagged change in purses was significant. Now, upon adding the lagged change in driving distance, we find that it is statistically significant. Thus, controlling for the lag structure of the first difference in purses upon itself, one finds that the lagged growth in driving distance has explanatory power. In this sense, there is evidence that driving distance Granger-causes purse size. From columns one and two it appears that driving distance is not influenced by purse size, but that the prodigious and increasing distance that golfers hit their tee shots does have the effect of increasing the prize funds, presumably through the avenue of higher gate and television revenues. Remember, of course, that these are marginal effects relevant for professional golfers. In an overall sense the purse size clearly effects driving distance. I play for purses of zero. Consequently, I rarely practice and do not hit my drives very far. I suppose that if there was enough money on the line I would take the time to practice more and could end up with longer tee shots than I currently have.

Table 8.3 Granger tests of purses and driving distance.

Descriptive statistics				
Variable	Mean	Standard Deviation	Minimum	Maximum
DRIVDIST[a]: men	275.6	10.01	260.2	286.9
DRIVDIST[a]: women	239.3	8.51	223.5	248.8
Purses[b]: PGA TOUR	167.9	88.5	49.4	279.4
Purses[b]: LPGA	38.2	12.0	20.4	63.2

Coefficient estimates (t-statistics)				
Sample	Men		Women	
Dependent Variable	ΔDRIVDIST	ΔPurse[c]	ΔDRIVDIST	ΔPurse[c]
Constant	0.27 (0.44)	6.86E+05 (0.39)	0.74 (1.04)	3.16E+05 (0.39)
$\Delta DRIVDIST_{-1}$		1.15E+06* (1.76)	0.19 (0.68)	3.76E+04 (0.12)
$\Delta DRIVDIST_{-2}$	0.47* (1.88)			
$\Delta Purse_{-1}$	7.8E-08 (1.01)	0.52** (2.57)	4.8E-07* (1.90)	7.81E-04 (0.00)
Adj. R^2	0.29	0.41	0.09	0.00
n	16	17	17	17

Notes: *, **, *** denote statistical significance at the 0.1, 0.05, and 0.01 levels
a. In yards
b. In undeflated millions of dollars
c. In deflated (1982-4) dollars

Interestingly, these inferences are reversed for women. In the third column of Table 8.3 the lagged change in purse size does influence the change in driving distance, even after controlling for the (insignificant) first lag in driving distance. So there is evidence consistent with the view that the increased prize funds for women have spurred efforts to increase the length of their tee shots, and that the efforts have paid off.

However, the increased length of women's tee shots does not seem to be a factor in the growth of prize money for women. In fact, there is virtually no explanatory power for the first difference in purses from either lagged changes in driving distance or lagged changes in the purses themselves.

Overall, these results are hampered by the small sample size. Unfortunately, the comparable statistics on golfer skills only go back so far. In my view, however, a coherent story is trying to emerge from the data. For men, the prodigiously increased length of tee shots, whether from technological innovation or hard work and practice, has been a factor in the fast-growing prize funds on the PGA TOUR. The "drive for show" part of the old adage is in operation, as the popularity of long drivers like Tiger Woods, Phil Mickelson, and John Daly fueled the growth in purses over this period. The same, however, is not true for women in LPGA events.

Meanwhile, on the incentives side, the purse growth does not seem to be a factor in bringing about longer drives for men. Perhaps men already are (and have been) maximizing their effort in this dimension of the game. Most of the increased length of drives could then be attributable to technology.[5] But for women, the increased purses may be bringing about extra practice and effort to increase driving distance in the desire to cash in on the higher prizes. Curiously, however, higher prizes have always been available to women if they wanted to compete directly against the men, but, historically, this did not bring the women's driving distance up to the men's level. However, as mentioned above, the relevant marginal increase in prize money would be that in LPGA events if that is where

[5]Technology growth may have been continuous over the period in question, but the application of the technology is also subject to the rules making bodies in professional golf. Interestingly, driving distance appears to have leveled after 2002, at about the same time when rules were adopted limiting certain technical aspects of golf ball and golf club design. See Stachura [2002].

women are typically competing. It is interesting that auxiliary regressions testing for Granger-causality from men's distance to women's purse growth or from men's purse growth to women's driving distance show no correlations whatsoever. Nor is there evidence that men's purses Granger-cause women's purses or that any women's variables Granger-cause any men's variables. As robustness and consistency checks on the reported regressions it is heartening that all of these latter nonsense regressions are insignificant.

8.4 Summary

The day when women can compete effectively against men in golf may not be coming soon. Most of the skills in golf are not necessarily dependent upon brute strength or physical size, but driving distance may be the exception. Therefore, if women's and men's skills converge in this dimension, then golf may become gender neutral. An examination of driving distance indicates that both men's and women's driving distances increased over the period of comparable data from 1992 to 2010, probably due to innovations in club and ball designs and materials. Unfortunately, although women made some inroads in catching up to men in the 1990's, the gains were transitory.

Although women and men compete separately in golf for the most part, this is not based on a *de jure* discriminatory segregation. Men are precluded from entering LPGA sponsored events but the PGA TOUR is explicitly non-discriminatory. The fact is that even though purses are lower in LPGA events, women self-segregate. Women find competition there only from other women and have higher expected returns, conditional on their skills, there than by attempting to join the PGA TOUR. Interestingly, women do seem to be motivated by prize levels in LPGA events to further develop the skill of driving distance, but do not seem to be motivated by the even higher purses on the PGA TOUR. Somewhat paradoxically, women are motivated by the attainable but lower prizes against lower competition in LPGA events more than they are by the much higher prizes on the PGA TOUR.

For their part, men do not seem to be motivated by increases in prize funds to increase their driving distances, at least not in the sense of Granger causality. Given the already high prize funds in 1992, perhaps men were already practicing to the point of sharply diminishing returns so that the extra motivation supplied by the increases in prizes did not lead to a measurably significant increase in practice. There does, however, seem to be Granger causality from driving distance to prize funds. Spectators enjoy with awe the prodigious distances that professional golfers can achieve. Statistically, this is one reason why purse increases have followed driving distance increases that are due to technological innovations in golf club and golf ball design.

Chapter 9

To Play or not to Play: The Skills Match[1]

Professional golfers make many choices over the course of their careers. For successful golfers who have achieved exempt status which guarantees entry into any tournament, the nature and extent of these choices actually increases. Golfers without exempt status or those struggling to keep it have their choices made for them in that they essentially have to try to enter every tournament they can. But golfers with exempt status can pick and choose how many and which tournaments to enter. This chapter and the following one report on my research examining the choice of which tournaments exempt golfers choose to enter. In this chapter I describe how a golfer who chooses to enter n tournaments in a particular season could optimally choose among the $N > n$ tournaments available to him, by choosing those in which he might expect to perform the best. Chapter 10 will follow up this analysis with a regression equation approach to the same problem.

Essentially, the choice of which tournaments to enter is part of what economists would call the labor supply decision of professional golfers. The labor supply issue in golf has been addressed from a variety of perspectives. Many economists [Shmanske, 1992, 2000, 2007, 2008, 2012;

[1]This chapter draws from material in Shmanske, S. (2009). Golf Match: The Choice by PGA Tour Golfers of Which Tournaments to Enter, *International Journal of Sports Finance*, 4(2), pp. 114-135. I would like to thank Leo Divinagracia for computer programming assistance, Kevin Quinn and Thomas Rhoads as discussants for an earlier version of this paper, conference participants at the May, 2008 IASE meetings in Gijon, Spain, and the July, 2008 WEAI meetings in Honolulu, and referees of *International Journal of Sports Finance*.

Moy and Liaw, 1998; Nero, 2001; Rishe, 2001; Alexander and Kern, 2005; Callan and Thomas, 2007] have estimated production functions wherein tournament scores are a function of the golfer's skills, or earnings functions wherein earnings are a function of the golfer's skills. Thus, the supply in different dimensions of talent leads to the performance or the earnings. Going one step backward in this supply chain, Shmanske [1992] also looked at the production, development, and maintenance of the skills themselves where the input supply is the golfer's practice time. Ehrenberg and Bognanno [1990a], [1990b], consider the golfer's supply of effort, especially in the final rounds of competition. Finally, Gilley and Chopin [2000], Hood [2006], and Rhoads [2007a], [2007b] study what might be called regular old supply, that is, the number of tournaments entered by professional golfers in a given year. The focus in these studies is on supply elasticity in order to determine whether a golfer's individual supply curve could be backward bending.

This chapter draws a little from each of these traditions but perhaps can be seen most closely as a complement to Rhoads's recent papers. Whereas Rhoads looks at the number of tournaments entered by a golfer, I hold that result constant and attempt to answer the question of which tournaments a given golfer will enter. In a nutshell, I do the following. First, by looking at the statistics from individual tournaments, I can determine which skills are the most important for success in each particular tournament. For example some courses will favor long drivers while at others accuracy is more important than length. Second, by looking at the statistics for each individual golfer over the course of a year, I can determine which skills are relatively weaker or stronger for each one. For example, some golfers are renowned for their prodigious length while others are known for accuracy or putting prowess. Third, by matching the skills that the golfer excels in to the skills more highly required in each tournament, I calculate an expected performance for each tournament. These expected performances are ranked from 1 to N, and if the golfer chooses to play in n of the N tournaments, the ideal set of which n tournaments the golfer should choose is determined. Finally, I compare the set of tournaments that a golfer actually enters to sets of tournaments chosen randomly to determine whether a golfer's choice is systematic or random with respect to the match between the skills required by a tournament and the skills at which the golfer excels.

Suppose a golfer wants to play in n of N tournaments. And suppose a golfer's expected performances in each of the N tournaments can be ranked from 1 to N. Then, the ideal choice is to pick those tournaments ranked 1 to n. This problem is equivalent to choosing n balls without replacement from an urn with N balls numbered 1 to N. If the balls are chosen randomly, combinatorial arithmetic reveals that the sum of the numbers on the balls chosen is distributed symmetrically around a mean of n(N + 1)/2 and ranging from $\sum_{i=1,n} i$ to $\sum_{i=N-n+1,N} i$. For example, in a small numbers illustration suppose a golfer wants to play in two of six possible tournaments. Then n = 2 and N = 6. The algebraic expressions indicate that the sum of the numbers on two balls chosen from an urn with six balls numbered 1 to 6 is distributed symmetrically around an average of 7 and ranging from a minimum of 3 to a maximum of 11. But all these possibilities are not equally likely. There is only one way to get a sum of 3, namely by choosing balls numbered 1 and 2, but there are a number of different combinations that yield a sum of 7. This example is extended later in the chapter. By summing the ranks of the tournaments actually chosen, it can be determined where in the random distribution this sum falls. If the sum is small enough to be in the lower tail of the actual distribution, then the hypothesis that the golfer chooses which tournaments to enter randomly with respect to the match of skills in hand and skills required can be rejected in favor of the hypothesis that the golfer is systematically choosing along these lines. There seems to be no other research exploiting combinatorial arithmetic in this way, and indeed, a computer algorithm had to be written in the course of this research.[2]

The ideal set of tournaments referred to in the aforementioned nutshell explanation is "ideal" in the limited sense of considering only skills and predicted scoring performance. There are many other factors that golfers consider. For example, a golfer might not be eligible for a tournament that suits his skills and style of play. Or, such a tournament might fall on the

[2]For a golfer choosing n=23 out of N=46 tournaments there are (46!)/(23!)(23!) different combinations. That is, there are more than a trillion combinations in the actual random distribution. With negligible difference, the research also uses a simulated distribution with a million trials.

weekend of the player's anniversary or other significant family event that makes the opportunity cost too high. Or, several potentially attractive tournaments might fall sequentially in a row and conflict with other priorities of the golfer such as not being away from home more than three weeks at a time. Additionally, there are three weeks during the season when two tournaments are scheduled and the golfer can only compete in one, even if both are attractive.

There are also dynamic considerations to consider. At the beginning of the year a golfer will typically map out his schedule of which tournaments he expects to enter. During the year, these entry decisions can change due to injury, due to an attempt to exploit a hot hand,[3] and most importantly, due to strategic entry at the end of the year in an attempt to cross thresholds in the yearly earnings lists. For example, ending the year in the top 30 on the earnings list allows entry into certain exclusive, high-purse tournaments with guaranteed payouts, while ending the year in the top 125 on the earnings list gives the golfer "exempt" status which allows entry into practically all of the official tournaments for the next year.[4] Without exempt status, a golfer cannot really choose which tournaments to enter, he typically will have to win a qualifying competition held early in the week of the tournament, or seek a sponsor's exemption. Any of these dynamic considerations, or idiosyncratic factors similar to those mentioned in the previous paragraph, introduce random noise into the calculation of the ideal set of tournaments, which biases the results against the rejection of the null hypothesis of random selection. If the golfer can still be shown to choose systematically with respect to the match between the skills in hand and the skills required, then the results are strong indeed.

[3]Connolly and Rendleman [2008] found evidence of statistically significant streaky play for 9% of the golfers in their sample.

[4]Typical tournament fields are 144 golfers, so exempt status from being in the top 125 assures entry. Tournament winners from previous years have a higher priority on the exempt list, but almost all of these winners are also in the top 125. Previous winners who are not also included in the top 125 include older golfers with "lifetime" exemptions such as Arnold Palmer and Jack Nicklaus, who will not enter current tournaments. The sample in this paper uses only the top 100 golfers of 2005, each of whom would have been eligible to enter any of the regular PGA TOUR events in 2006.

As described so far, the golfer chooses only in terms of his ranking of expected performance in terms of final score. But expected score is an indirect proxy for earnings for at least two reasons. First, the purses differ by tournament. A golfer may choose a tournament where he has a higher expected score (higher scores being worse in golf) if the purse is sufficiently larger. Second, the tournaments differ by the strength of the competition. A golfer may choose a tournament where he has a higher expected score because fewer of the elite golfers choose to enter such that the higher expected score translates into a better absolute rank in the tournament, and therefore, higher earnings.

The preceding arguments suggest a refinement of the original, nutshell, explanation. Instead of ranking all tournaments by expected scoring performance, one should rank only those tournaments for which one is eligible. This approach should yield better results than by only looking at expected score, but a further refinement may also be possible. Instead of ranking the tournaments for each golfer based on expected score, they could be ranked based on expected earnings. This is accomplished by comparing the expected score of each golfer to the expected scores of all the golfers who enter a particular tournament to calculate an expected rank finish order for each golfer. Combining information about the expected rank finish order with information about the structure of the prizes leads to a point estimate of the expected earnings for each tournament and an alternate ranking of the ideal set of tournaments.

Finally, a simple ranking of the tournaments by the size of the purse indicates how important the prize fund is to alluring the top talent.

The strength of the combinatorial analysis carried out in this chapter is that it imposes very little structure on the golfer's choice. The weakness, however, is that without *ad hoc* weighting, only one dimension can be used at a time to achieve the ideal ranking of the tournaments. Multiple regression analysis using a logistic transformation can simultaneously assess the effects of the purse, the strength of the competition, and the skills match, at the cost of imposing significant structure on the model. Furthermore, regression analysis can also simultaneously control for the dynamic considerations discussed above, such as, hot hand status, injury, or year-end strategic entry to maintain exempt status for the following year. A regression analysis of the entry choice is covered in the next chapter.

The remainder of this chapter has two main parts. The data collection and manipulation is described in the next section. This section develops the rankings of the tournaments for each golfer. The statistical hypothesis tests using the combinatorial distribution of the sum of the ranks for the tournaments chosen is presented in the section following that.

9.1 The Data

The PGA TOUR sponsored 48 "official" events during the 2006 season. After each tournament the PGA TOUR updates its website (www.pgatour.com) and reports the year-to-date performance statistics of the golfers. Although the PGA TOUR would not share its data, it is possible to back out the weekly performance statistics from the change in the year-to-date statistics for a pre-chosen set of 100 top golfers.[5] Two of the forty-eight tournaments were not used because they used alternative scoring formats in which golfers play different numbers of holes and not every stroke counts. Not every player plays in every tournament so of the 4600 possible tournament entries, 2,360 are actual entries for which the performance statistics and scoring outcomes are observed.

For each of these 2,360 observations, six statistics were tracked: The score per 18 holes measured in strokes, SCORE; the driving distance measured in yards, DRIVDIST; the driving accuracy measured as the percentage of drives ending in the fairway, DRIVACC; the approach shot accuracy measured as the percentage of greens reached in regulation, GIR; the putting proficiency measured as the number of putts taken per green reached in regulation, PUTTPER; and the sand bunker skills measured as the percentage of times two or fewer strokes are taken to finish a hole from a greenside bunker, SANDSAVE. Summary statistics for these, their transformations, and the other variables used appear in Table 9.1.

[5]The study tracked the top 100 money winners of 2005 throughout the 2006 season. These golfers accounted for the lion's share of the total money awarded in 2005 and again in 2006. The sample size is arbitrary but is large enough to give ample degrees of freedom for the tests performed in this and other research stemming from the sample. This is the same data that formed the basis of Chapter 6 and is explained in more detail there.

Table 9.1 Summary statistics.

Variable	Mean	Std. Dev.	Minimum	Maximum	N
SCORE	71.46	2.32	65	82	2360
DRIVDIST	289.09	13.90	184.4	353.2	2360
DRIVACC	0.636	0.120	0	1	2360
GIR	0.655	0.083	0.38	1	2360
PUTTPER	1.777	0.090	1.403	2.133	2360
SANDSAVE	0.496	0.207	0	1	2360
ASCORE	3.024	1.874	-3.793	11.996	2360
ADRIVDIST	-18.052	10.347	-120.241	32.324	2360
ADRIVACC	0.008	0.095	-0.664	0.605	2360
AGIR	-0.102	0.068	-0.384	0.409	2360
APUTTPER	0.031	0.082	-0.289	0.346	2360
ASANDSAVE	-0.050	0.201	-0.675	0.537	2360

9.1.1 *Calculating the golfer's skill set*

The observed measures of these skills, and the scores themselves, are not directly comparable across tournaments because the conditions at each tournament differ with respect to weather, course conditions, length of course, elevation above sea level, size of greens, width of fairways, number of trees, lakes, and other hazards, etc. Therefore, six models of the following form are estimated:

$$X_{ij} = CB_C + GB_G + E_{ij} \qquad 1 \leq i \leq 46 \text{ and } 1 \leq j \leq 100 . \qquad (9.1)$$

In Eq. (9.1), X_{ij} is the vector containing the dependent variable, SCORE, DRIVDIST, etc., with element x_{ij} capturing the performance observed on the ith course by the jth golfer. For example, for the SCORE equation, x_{ij} is the average score per round for the jth golfer in the ith tournament. C is a matrix of dummy variables, one for each course. That is, it has a one in the ith place and zeroes elsewhere to separate out the effect of each course on the final scores. Thus, B_C is the vector of 46 coefficients controlling for the average SCORE in each tournament. G is similar to C in that it is a matrix of dummy variables, but this time there is one dummy variable for each golfer. Thus, G has a one in the jth place and zeroes elsewhere to capture the average effect of the jth golfer. So, corresponding to G, B_G is the vector of 99 coefficients (one for each golfer with Tiger Woods omitted to avoid singularity in the estimation process) to control for the fact that golfers with different sets of skills play in different tournaments. The convention of leaving out one golfer means that the individual elements in the B_G vector capture the difference between each particular golfer's SCORE (or DRIVDIST, etc) and that of Tiger Woods. Finally, E_{ij}, with individual element, e_{ij}, is the vector of error terms. Thus, the estimates in B_G give the average levels of each of the five skills by golfer, and the average scores by golfer, both relative to Tiger Woods, controlling for the skills of the other golfers and the difficulty or ease of the courses at each tournament's venue.

Rewriting Eq. (9.1) yields:

$$X_{ij} - CB_C = GB_G + E_{ij} \equiv AX_{ij} \qquad 1 \leq i \leq 46 \text{ and } 1 \leq j \leq 100 . \qquad (9.2)$$

This defines a vector, AX_{ij}, with individual element, ax_{ij}, of the measurements of skills and scores adjusted for course effects and relative to Tiger Woods. That is, AX stands for ASCORE, ADRIVDIST, etc., which in turn stand for the adjusted scores, adjusted driving distances, etc., and as Eq. (9.2) shows this can be interpreted in two ways. First, AX can be seen as the actual skill level or score, x_{ij}, minus a course adjustment factor given by the appropriate element of B_C. Second, AX can be seen as the average level of adjusted skills or scores for each golfer, given by the appropriate element of B_G, plus an error term for that golfer in that tournament.

The coefficient estimates in B_G are the results to take away from this exercise. They capture the average skill level in each dimension for each golfer. These skill levels are used in calculating a prediction for how well a golfer will do in any particular tournament. These estimates will be referred to as the golfer's skill set.

9.1.2 *Calculating the required skills by tournament*

To estimate how well a golfer will do in any tournament we also need the relationship between the skills and scores on a tournament-by-tournament basis. Thus, the following equation is estimated individually on data for each tournament:

$$\begin{aligned} \text{SCORE} = \; & a_0 + a_1\text{DRIVDIST} + a_2\text{DRIVACC} + a_3\text{GIR} + \\ & a_4\text{PUTTPER} + a_5\text{SANDSAVE} + e\,. \end{aligned} \tag{9.3}$$

Eq. (9.3) will tell us what effect each of the five skills has on the final score on a tournament-by-tournament basis. One way to get a feel for the importance of the right hand side variables relative to each other is to estimate Eq. (9.3) as a standardized regression, in which, essentially, the variables are converted to z-scores from normal distributions. For example, an increase of one percentage point in approach shot accuracy as measured in GIR, should be associated with lower scores, and will be measured in the estimate of a_3 in a normal regression. But there is no easy way to tell whether a one percentage point increase is a lot or a little. In a standardized regression, the estimate of a_3 gives the relationship between scores and approach shot accuracy where the latter is measured in standard deviations above the mean for the particular skill in question. Now we have something to gauge by. In general, moving from the mean to one standard deviation above the mean means leapfrogging about one third of the other golfers in terms of the skill. To illustrate the effect, Table 9.2 lists the results for one of the 46 tournaments.

In Table 9.2, The Honda Classic is chosen as an illustration because it shows that neither the t-statistics from the Ordinary Least Squares (OLS) regression, nor the magnitudes of the coefficients in the OLS regression give the give the correct ranking of the importance of each of the skills on

Table 9.2 Estimates (t-statistics) of Eq. (9.3) for Honda Classic, March 9-12, 2006.

Equation	9.3	9.3
Method	O.L.S.	Standardized Regression
Constant	75.860*** (10.86)	1.293*** (6.01)
DRIVDIST	- 0.046** (-2.56)	-0.578** (-2.56)
DRIVACC	-6.716** (-2.37)	-0.558** (-2.37)
GIR	-11.606** (-2.48)	-0.587** (-2.48)
PUTTPER	12.687*** (5.24)	1.157*** (5.24)
SANDSAVE	-1.903* (-1.78)	-0.392* (-1.78)
adj. R^2	0.57	0.57
n	46	46

Note: *, **, and *** indicate significance at the 10%, 5%, and 1% levels.

the right hand side.[6] First, note that the linear transformation involved in reformulating the independent variables do not change the statistical

[6]Neither do the implicit elasticities. Elasticities measure sensitivity based on the average magnitude of each independent variable, whereas the coefficient estimates from the standardized regression measure sensitivity based on each golfer's placement in the distribution of the skill taken over all golfers and measured in standard deviations. For example, consider driving distance in which a 30-yard improvement might be only a 10% change when calculated in an elasticity if average drives are 300 yards, but may represent two or three standard deviations above the mean and be very difficult to achieve.

properties of the regression. Neither the t-statistics nor the overall R^2 change. In the standardized regression in Table 9.2 a one standard deviation increase in PUTTPER (that is, a decrease in skill) increases one's score by over 1.1 strokes per round. Meanwhile, one standard deviation increases in DRIVDIST, DRIVACC, or GIR will save the golfer over one-half of a stroke per round on average. In this sense, putting is clearly the most important skill for achieving success in this tournament. The correct ranking of the order of importance of the skills is putting, first, which is roughly twice as important as the closely grouped greens in regulation, driving distance, and driving accuracy, in that order, and followed by sand saves.

The standardized regressions were carried out for all 46 of the tournaments. Out of 230 (5 times 46) coefficient estimates, 138 are statistically significant with the correct sign, one is statistically significant with the incorrect sign, and 92 are insignificant. With one exception, the adjusted R^2 ranges from 0.27 to 0.84 and the sample size ranges from 22 to 96.[7] Table 9.3 captures the frequencies with which each pattern of statistically significant estimates shows up in the 46 tournaments. Putting is the most important skill in 25 of the 46 tournaments and second most important in 19 others. Reaching greens in regulation (GIR) is the most important in 20 of the tournaments, and second most important in 21 others. DRIVDIST is important in 23 tournaments. DRIVACC is important in 15 tournaments and SANDSAVE is important in 23 tournaments.

The rank order pattern of the importance of the different skills illustrates that different tournaments differentially reward different skill sets. This is a necessary step in the analysis because if the pattern and magnitude of the importance of the skills was the same in each tournament,

[7]The exception is the U.S. Open which returned no significant coefficients and an adjusted R^2 of zero. The U.S. Open seems to bamboozle statistical researchers as much as it does professional golfers. This result should be of concern to the United States Golf Association (USGA) which runs the U.S. open. While the USGA prides itself on setting up a tough golf course for the tournament, these results beg the question of what golf skills the USGA is trying to reward. Evidently, neither the ability to drive the ball long and straight, nor the ability to hit greens with approach shots and make putts, nor the ability to recover from greenside bunkers are correlated with lower scores. And collectively, these skills explain none of the variation in final scores for the U.S. Open whereas they typically explain one half or more of the variation.

Table 9.3 Frequencies of patterns of importance.

Pattern, and order of importance					
First	Second	Third	Fourth	Fifth	Frequency
PUTTPER	GIR	DRIVDIST	DRIVACC	SANDSAVE	1
PUTTPER	GIR	SANDSAVE	DRIVACC	DRIVDIST	1
GIR	PUTTPER	DRIVDIST	SANDSAVE	DRIVACC	1
PUTTPER	GIR	DRIVACC	SANDSAVE		2
PUTTPER	GIR	SANDSAVE	DRIVACC		2
PUTTPER	GIR	SANDSAVE	DRIVDIST		1
PUTTPER	GIR	DRIVDIST	SANDSAVE		1
PUTTPER	DRIVDIST	DRIVACC	GIR		1
GIR	PUTTPER	DRIVACC	DRIVDIST		1
GIR	PUTTPER	SANDSAVE	DRIVDIST		1
GIR	PUTTPER	SANDSAVE			10
GIR	PUTTPER	DRIVACC			2
PUTTPER	GIR	SANDSAVE			2
PUTTPER	GIR	DRIVDIST			2
PUTTPER	GIR	DRIVACC			2
PUTTPER	DRIVACC	GIR			1
PUTTPER	SANDSAVE	GIR			1
GIR	PUTTPER	DRIVDIST			1
GIR	DRIVACC	PUTTPER			1
PUTTPER	GIR				7
GIR	PUTTPER				3
PUTTPER					1
none					1

or statistically insignificant, then golfers would not be able to even attempt to choose tournaments based upon a match between their skills and the skills rewarded at that golf course. But as we can see, there is a difference and golfers could choose along these lines.

The rank ordering of the importance of the skills shows that there are differences, but loses the continuous nature of the data. For example, PUTTPER and GIR might be ranked first and second most important, and be roughly equal in importance, or PUTTPER could be much more important to the outcome. For this reason, the actual coefficient estimates in Eq. (9.3) will be used as weighting factors for the skills necessary in each tournament.

9.1.3 *Calculating the ideal set of tournaments for each golfer*

The identification of the ideal set of tournaments for each golfer follows directly. Using Eq. (9.3) and plugging in the skill sets identified for each golfer, yields the predicted relative score of each of the 100 golfers for each of the 46 tournaments, 4,600 predictions in all. These relative scores are grouped by golfer to yield a 1-46 ranking of the tournaments for each golfer. A golfer choosing to play in n of the 46 tournaments, should choose the top ranking n. A primary goal of the research in this chapter is to determine how well the golfers make this choice. To do this, we must find out where the tournaments actually chosen by a golfer place in the distribution of all possible choices.

If the golfers choose which tournaments to enter based on the effects that are calculated in this chapter, then comparing the golfer's actual choice to a choice made randomly among the 46 available tournaments should indicate it. Random choice among the 46 tournaments is equivalent to choosing n balls without replacement from an urn containing balls numbered 1 to 46. Ultimately, this chapter will address the problem by comparing the average of the numbers on the n chosen balls to the probability distribution of the average taken on n balls chosen randomly.

Consider the following small numbers illustration of the problem. One golfer wants to choose two tournaments from a set of six available tournaments which a prediction model like that explained above has ranked from one (best) to six (worst). Another golfer wants to play in three tournaments. The best choice for the first golfer is the set of tournaments characterized by the pair (1, 2). The worst choice would be (5, 6). There are 15 different combinations in all. For the second golfer the best choice is (1, 2, 3), the worst choice is (4, 5, 6), and there are 20 different combinations. To compare the choices of the golfers to random choice first consider the probability distribution function of the sum of the ranks. For the first golfer this appears in the top panel of Fig. 9.1. For the second golfer this appears in the middle panel of Fig. 9.1. These distributions of the sum of the numbers on n balls chosen without replacement will always be symmetric around a mean of $n(N+1)/2$. Thus, as n increases, the distribution slides to the right. Accordingly, the distributions can all be "centered" around a mean of $(N+1)/2$ by dividing the sum of the numbers on the balls by n. This is done and the distributions overlapped in the bottom panel of Fig. 9.1.

Continuing with the small numbers example, suppose the first golfer actually chose the set of tournaments given by (1, 2). The average rank for his choice is 1.5 and is located in the left tail of the distribution with a probability of being that low (or lower) of one-fifteenth. That is, it would happen randomly only one-fifteenth or about 6.67 percent of the time. The p-value associated with this outcome is simply this probability, in this case 0.067. Later, the p-values achieved by each of our 100 golfers will be highlighted. Thus, the null hypothesis of random choice for this golfer would be rejected (at the 10% level because the p-value is less than 0.10, but not at the 5% level) in favor of the alternative hypothesis that the golfer chooses systematically, as ranked by the model, which tournaments to enter.

The small numbers example provides the blueprint for what will be done for each of the 100 golfers in the sample, except the golfers choose anywhere between 4 and 32 tournaments out of 46 possible tournaments. As such, the distributions of the averages of the ranks of the tournaments chosen will be centered around 23.5. These distributions get very big. As an example, consider the choice of 15 tournaments out of 46. There are over 517 billion ($46!/15!31!$) different ways to choose them. Fig. 9.2 shows the left tail of this distribution. The lowest possible sum is 120 if the golfer chose tournaments ranked 1 through 15. This would lead to an average

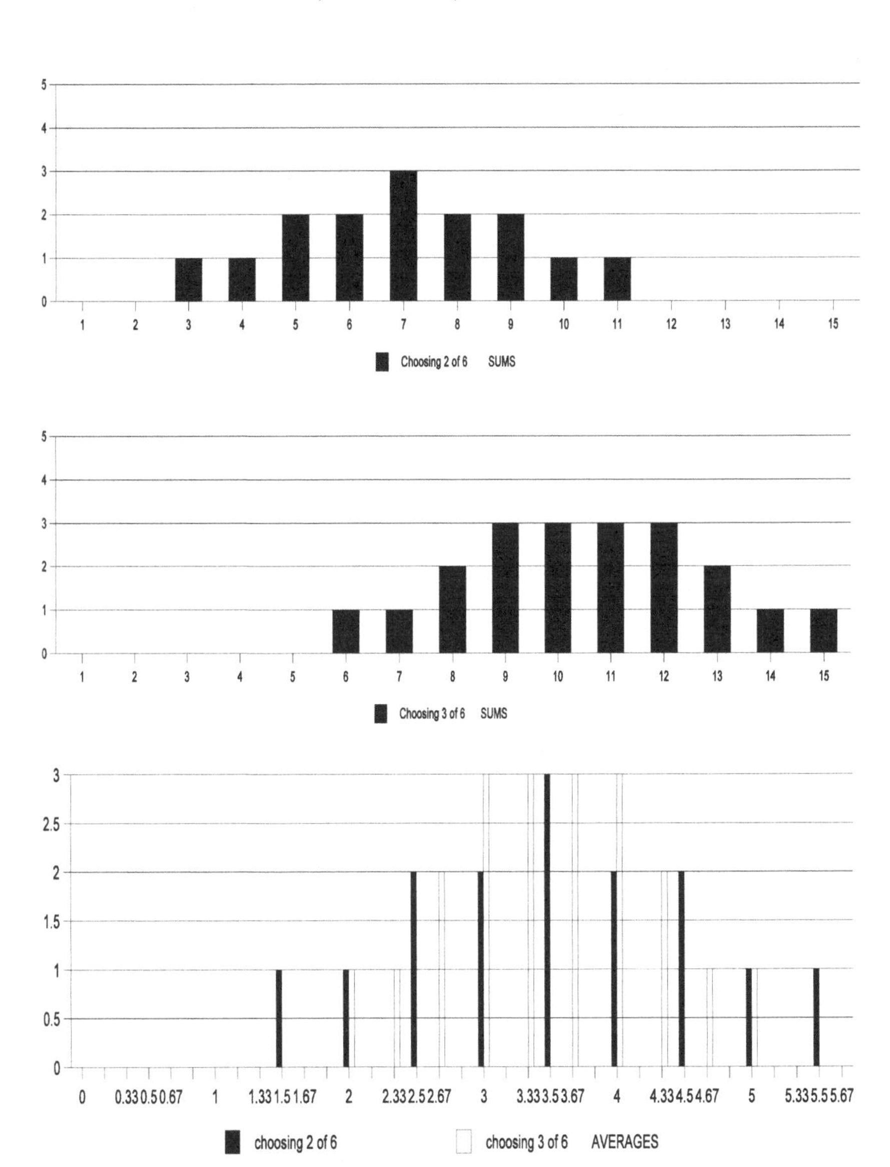

Fig. 9.1 Distributions of sums or averages of ranks.

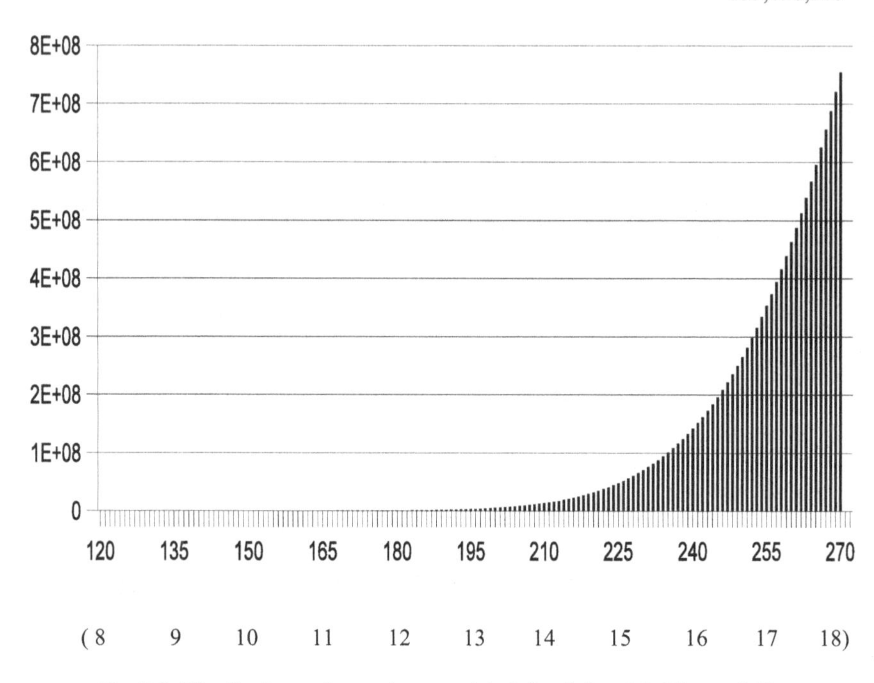

Fig. 9.2 Distributions of sums (averages) in left tail for pick 15 out of 46.

rank of 8, that is, 120/15. And there is only one combination that yields this sum. Now consider a sum of 270 for an average rank of 18. There are over 755 million combinations that add up to a sum of 270, but this is still tiny compared to the total number of possibilities.[8] The cumulative probability of an average rank no larger than 18 is still less than 2.7%. Therefore, if a particular golfer chose to play in 15 tournaments, and based on the ranking in this paper chose tournaments with an average rank of 18,

[8]A computer program was designed to count the number of different combinations leading to different sums or averages in the lower tails of the overall distributions. To count all the combinations leading to sums no larger than 270 (averages up to 18) for the choose-15-out-of-46 problem took five days on my desktop computer. To count out the whole distribution would take much longer. As n increases from 15 to 23 (the largest problem) the counting goes up by an order of magnitude for each step of n. For the largest distributions, the probability densities and p-values are estimated with only negligible error in what is called a "simulated bootstrap distribution" based on a million random samples.

we would reject the null hypothesis (with a p-value of .027) of random tournament selection in favor of the alternative hypothesis of systematic tournament choice as described by the model. Indeed, one such golfer in the sample, Mark Hensby, is precisely described in this example.

9.2 Results from the Combinatorial Analysis

It would be awkward to enumerate the results for each of the 100 golfers. Instead, the distributions will be summarized in a bar graph. Some summary statistics will be presented and they will show the extent to which the golfers choose systematically based on matching the skills they have with the skills required at different tournament sites. In Fig. 9.3 the solid

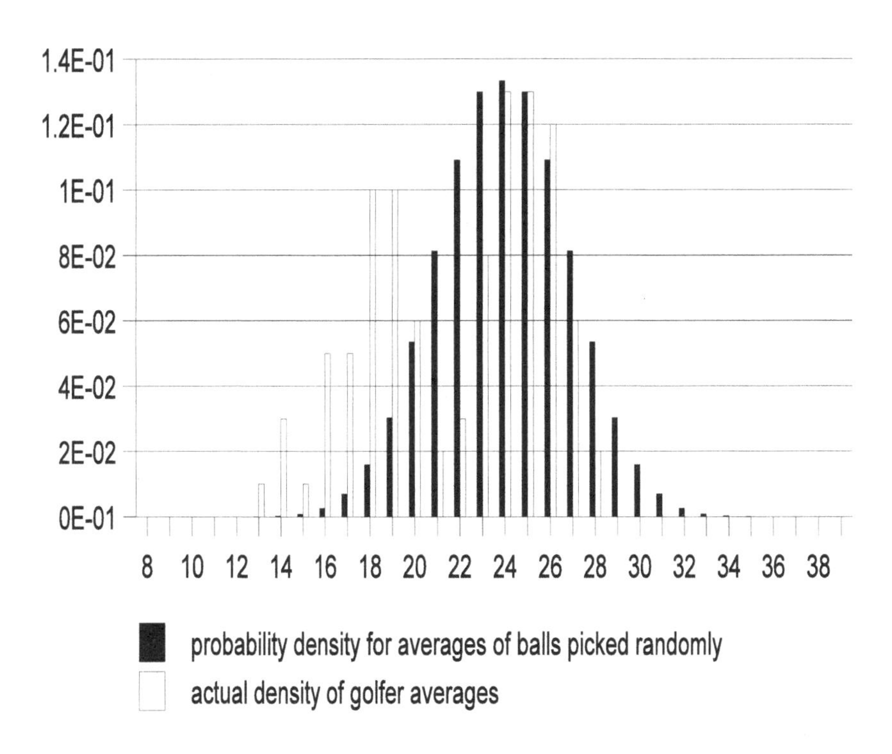

Fig. 9.3 Comparison of random probability density with density of actual golfer averages.

bars show the probability density function for the average of the pick 15 balls out of 46 problem. The lowest average possible is 8, the highest 39, and the midpoint is 23.5. As such, Fig. 9.3 gives a complete picture of the probability distribution function, the left tail of which was pictured in Fig. 9.2, in the following manner. The solid bar at 18 in Fig. 9.3 depicts the proportion of all the 517+ billion possibilities that have averages falling in the range (17, 18], that is, all the cases above 256 and up to and including 270 in Fig. 9.2. There are a lot of cases in this interval, over 755 million at 270 alone, but as a proportion of the total, they make up less than 2% of the total. The solid bar at 24 covers the interval from 23 to 24 and as such, includes the mean at 23.5. The critical values at the 1%, 5%, and 10% probability levels fall at the averages 16.93, 18.8, and 19.8, and therefore, fall in the bars at 17, 19, and 20, respectively.

Now consider the averages achieved by the 100 golfers in the sample as depicted by the density function with the striped bars. A significant number of golfers clearly appear to pick their tournaments systematically according to the predictions in the model. A full 45% of the golfers achieve averages in the 10% tail of the random distribution. The distribution of the golfer averages appears to be bimodal, with one group being able to achieve an average in the 17 to 19 range and another group bunched in the 23 to 25 range. For the group clustered in the bars at 24 and 25, the null hypothesis that they are choosing from among the 46 tournaments randomly with respect to the skills match as described in this paper cannot be rejected. However, for those clustered in the bars at 19, 18, and below, the null hypothesis of random tournament entry is clearly rejected.

Figure 9.3 is illustrative, but does not give a fully accurate depiction of the probability levels involved because the critical values are exactly calculated only for the choose 15 out of 46 tournaments problem. But in the sample of 100 golfers, one chose only four and several chose as many as 32. Figure 9.1 illustrated for a small numbers case how choosing more tournaments cut out the extreme tails and pushed more density toward the mean. The same tendency exists for the large numbers problem. So, instead of a figure illustrating the density for each choice of n out of 46, each golfer will simply be placed in the proper distribution for however many tournaments he chose and the probability value for rejection of the

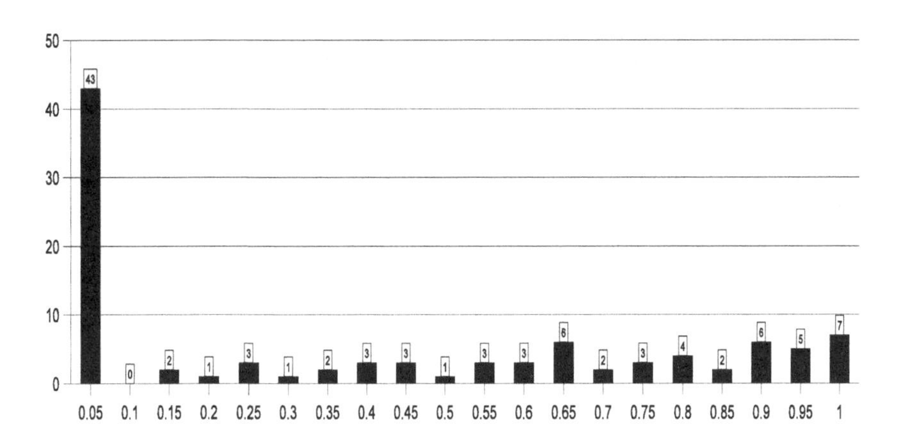

Distributions of p-values (all 46 tournaments)

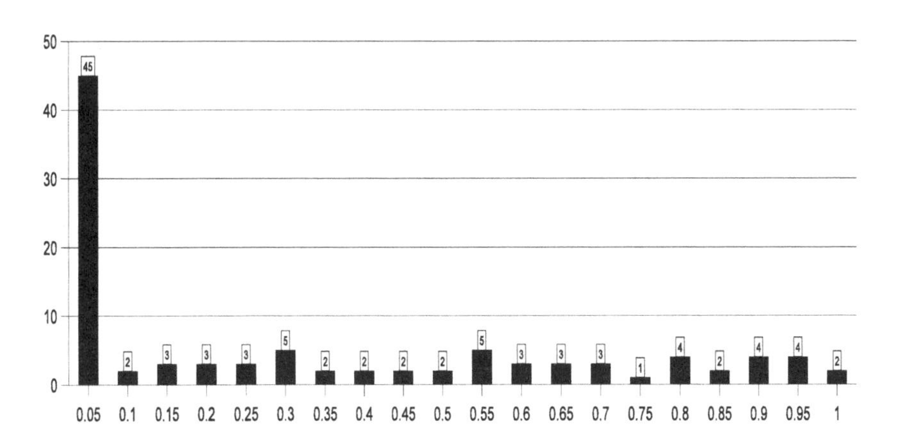

Distributions of P-values (ineligible-player-tournaments excluded)

Notes: The bar at 0.05 indicates all those whose average ranks have p-values less than 0.05 ($p < 0.05$), the bar at 0.1 indicates all those whose average ranks have p-values from 0.05 up to but not including 0.1 ($0.05 \leq p < 0.1$), and so on.

Fig. 9.4 Distributions of p-values.

null hypothesis is calculated. A bar graph showing the distribution of these p-values is depicted in Fig. 9.4.

The top panel in Fig. 9.4 is the first main result of the research. If the golfers chose their tournaments randomly with respect to the factors considered here one would expect there to be about five golfers in each of the 20 intervals formed when the probability level on the horizontal axis is divided into segments of length equal to 0.05. This clearly does not happen. A full 43% of the golfers fall into the less than 5% range and one can confidently reject the null hypothesis of random selection for these. Furthermore, of these 43 golfers, the null can be rejected at the 1% level for 30 of them. Meanwhile, the other 57 golfers are spread pretty evenly throughout the rest of the distribution.

The results in the top panel of Fig. 9.4 can be improved slightly. Not all golfers are eligible for all tournaments. In particular, the season-ending TOUR Championship, the season-opening Mercedes Championship, and the famous Masters tournament have limited entry fields that deny entry to a portion of our sample. Furthermore, during some weeks there are two official tournaments and a golfer could choose at most one. Instead of ranking all 46 tournaments, golfers will be choosing out of 43, 42, or fewer tournaments. The appropriate adjustments are made and new p-values are calculated for each golfer's choice of tournaments and the results charted in the bottom panel of Fig. 9.4. As shown, the results improved slightly. The null is rejected at the 5% level for 45 golfers, and for two more at the 10% level. Indeed, 37% of the golfers even fall into the 1% tail of the appropriate probability distribution.

As outlined in the introduction, an additional refinement is to rank the tournaments on the basis of expected earnings as opposed to expected relative scoring performance. This exercise brings two more pieces of information into the decision, the size of the purse and the strength of the competition. Each golfer's expected relative score at a tournament, as calculated above, is now compared to the expected scores of the other top golfers who enter that same tournament to return an expected rank-order of finish for each tournament. Coupled with the size of the total purse and the nonlinear distribution of the individual prizes,[9] the expected earnings in each tournament can be calculated and will form the basis for each golfer's

[9]For the vast majority of tournaments first prize wins 18% of the purse, second place wins 10.8%, and so on down to 0.2% for 70th place.

ranking of the tournaments. The results are pictured in the top panel of Fig. 9.5.

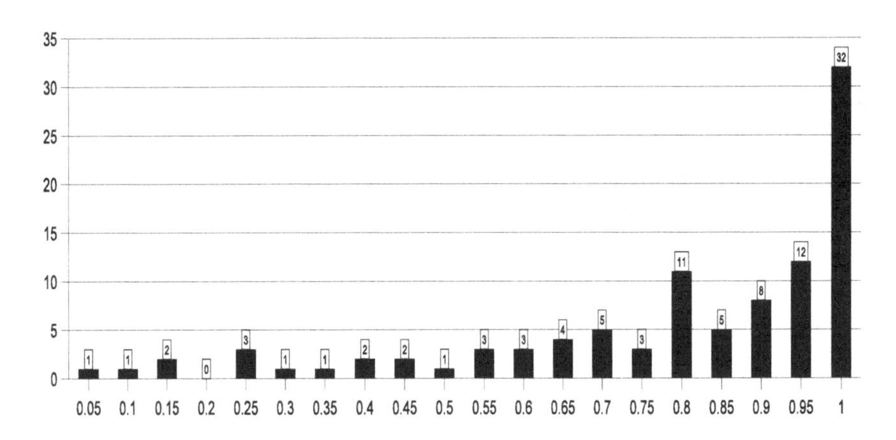

Distributions of p-values (ranks based on expected prize)

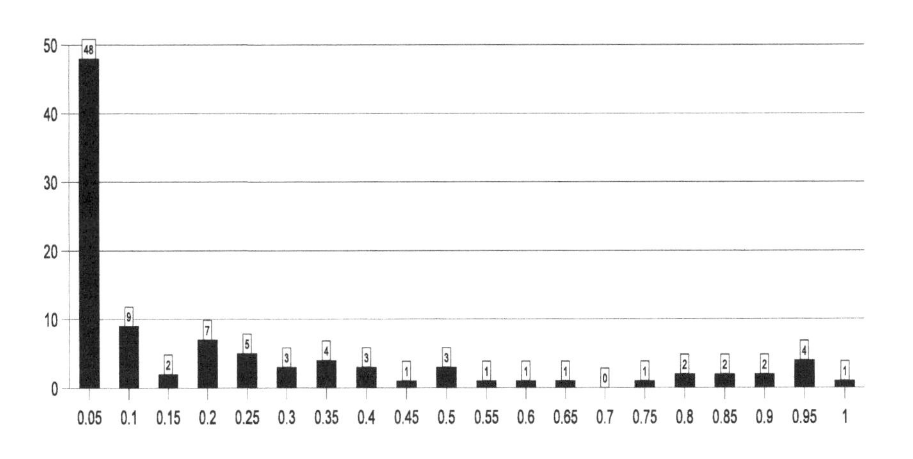

Distributions of p-values (ranks based on total purse, then expected relative score)

Notes: The bar at 0.05 indicates all those whose average ranks have p-values less than 0.05 ($p < 0.05$), the bar at 0.1 indicates all those whose average ranks have p-values from 0.05 up to but not including 0.1 ($0.05 \leq p < 0.1$), and so on.

Fig. 9.5 Distributions of p-values.

As the top panel of Fig. 9.5 shows, there is no support for the proposition that a golfer chooses which tournaments to enter based on his expected prize as calculated in this model. If anything, there is support for the opposite. For many of the golfers the null hypothesis of random selection would be rejected in favor of the hypothesis that golfers chose tournaments based on the *reverse* order of expected earnings as calculated here. As the bottom panel will confirm, the problem is not with including information about the purse, but rather with the inclusion of the strength of field information. Indeed, for many of the major tournaments and the tournaments with the highest purses,[10] more than 70 of the golfers in the data competed, even though the prize funds distribute money only down to 70th place. This means that all golfers with expected finish ranks of worse than 70 would actually "expect" a prize of zero by the calculations here. If these golfers choose based on expected prize, they would rank these tournaments last among all tournaments, yet they did choose to compete. Perhaps these golfers are overly optimistic about their chances against all of the other top golfers, but one should not discount the golfer's belief in his own skill and in the chance that he can turn in an above average performance. There may also be non-pecuniary reasons for competing in these tournaments such as fame and fortune coming from winning a major, or such as future entry into other restricted tournaments. Whatever the reason, many tournaments are entered even when the expected prize is smaller than the expected prize for other tournaments that were skipped.

The bottom panel of Fig. 9.5 shows the distribution of p-values for the average rank of the tournaments chosen when tournaments are ranked looking first only at the total purse, and looking second (as a tie-breaker) at the expected performance in the tournament based on the skills the golfer possesses and the skills required by the tournament venue. That is, golfers rank which tournaments to enter based on the total purse, but in the case where two tournaments have the same purse, the golfer ranks them by expected performance as above in Fig. 9.4. Thus, this measure ignores the strength of the competition. There is strong support for this decision rule, with close to half the sample rejecting the null of random selection in favor of the ranking based predominantly on total purse.

[10]Overall, purses ranged from $3 million to $8 million.

As is often the case with research, obtaining an answer to one question brings forth another question. The above histograms of p-values indicate that roughly half of the 100 professional golfers studied in this research make their tournament entry decisions systematically with respect to the purse size and the match of their skills with the skills that are rewarded at each particular tournament venue. But what about the other half?

I compared the set of golfers for whom the null hypothesis of random selection is rejected in favor of the alternate hypotheses of the ranking based on the skills match (Fig. 9.4) or the purse (Fig. 9.5 bottom panel) to the set of golfers for whom the null is not rejected. The comparison was largely uninformative. There was no significant difference in the means between the two sets for age (36.4 to 36.9 years), experience (157 to 160 career tournaments), earnings (1.56 to 1.46 million dollars) or earnings per tournament (62 to 57 thousand dollars). There was, however, a significant difference[11] in the mean number of tournaments entered with golfers choosing wisely with respect to the skills match also choosing fewer tournaments in all (22 versus 26).

The fact that one cannot separate the "randomly-choosing" golfers from the "systematically-choosing" golfers by looking at previous earnings, or earnings per tournament, does not mean that the ability to systematically choose has no value. The value of correct choice is masked when looking backward at uncontrolled means of the two sets of golfers. A golfer could be in the top 100 money winners because he has superior skills, or because, given adequate skills, he chooses the correct tournaments, (or both). Once controlling for skills in a multiple regression, the independent effect of correct entry choice shows up. So, consider the regression of the logarithm of earnings per tournament[12] (LnEARNPERT) on the adjusted skills and the p-value from the calculations underlying the bar chart in the bottom panel of Fig. 9.4.

[11]The p-value was less than 0.01.

[12]As is typical, using the natural logarithm of earnings corrects for heteroskedasticity which is a problem when levels are used in an earnings equation.

$$\begin{aligned} \text{LnEARNPERT} = \; & \underset{(79.9)}{11.35} + \underset{(3.34)}{0.0317}\ \text{ADRIVDIST} + \\ & \underset{(2.42)}{3.66}\ \text{ADRIVACC} + \underset{(3.57)}{7.31}\ \text{AGIR} - \\ & \underset{(-5.61)}{11.68}\ \text{APUTTPER} + \underset{(2.65)}{2.75}\ \text{ASANDSAVE} - \\ & \underset{(-4.02)}{0.713}\ \text{P-VALUE} + e. \end{aligned} \tag{9.4}$$

n = 100, Adj. R-square = 0.658 (t-statistics)

The skills variables are averages, grouped by golfer, of the adjusted skills as defined in Eq. (9.2). The dependent variable is the logarithm of official earnings per tournament in 2006. All the skills are statistically significant in the expected direction. The variable of interest, P-VALUE, is low when the golfer is systematically choosing correctly, and therefore, its significant negative coefficient means that correct choice increases earnings as is expected. It is clear that the ability to more correctly choose which tournaments to enter will lead to higher earnings. But it remains a puzzle why some golfers are able to systematically choose the best tournaments for their skill set while others are not able to do so.

Chapter 10

To Enter or not to Enter: Another Look[1]

The previous chapter looked at a golfer's decision about which tournaments to enter as a problem in combinatorial arithmetic. As the reader might have gathered from the background explanation involved, this method is not the usual one that economists employ. Chapter 10 will use a more familiar method, regression analysis, to examine how golfers decide which tournaments to enter.

The previous chapter started by determining a ranking of expected outcome by golfer for each possible tournament. This was achieved in two steps. First, by looking at the correlation between scoring and golf skills in each individual tournament it was quantitatively determined how important each skill is at each tournament venue. Second, by tracking each golfer's performances over the course of the season we determined how much of each skill each golfer possesses. Then, simply put, golfers who are longer drivers than average ought to enter those tournaments that reward the skill of driving distance most handsomely, and this goes for each of the skills considered simultaneously.

However, there are other things besides the match between the skills the golfer has and the skills required by the tournament, so that a ranking of which tournaments to enter based only on the match is incomplete. This

[1]This chapter and the preceding one draw from material in Shmanske, S. (2009). Golf Match: The Choice by PGA Tour Golfers of Which Tournaments to Enter, *International Journal of Sports Finance*, 4(2), pp. 114-135.

chapter uses multiple regression analysis to simultaneously consider the match, as described above, and other factors that can be quantified.

But first, to solidify the link between the two chapters, look only at the match and consider the simple regression model capturing the following relationship:

$$\text{Probability of entering a tournament} = f(\text{constant}, \text{EXPECTED RELATIVE SCORE})\,. \quad (10.1)$$

As the EXPECTED RELATIVE SCORE goes up, remembering that higher scores are worse in golf, the probability that a golfer would choose to enter that tournament should go down. To address each golfer's entry decisions, Eq. (10.1) was estimated for each of the 100 golfers using binomial logistic regression.[2] The independent variable, EXPECTED RELATIVE SCORE, was developed in the last chapter and formed the basis for the ideal rankings that were used in the combinatorial analysis. Since higher scores are worse in golf we expect a negative coefficient on the expected relative score variable.

At this point, there are two important differences between the combinatorial procedure and the regression analysis. First, the regression analysis has the advantage of being able to use the continuously-measured actual levels of the expected relative scores instead of the 1-46 integer rankings. For example, suppose the 22nd and 23rd ranked tournaments for a particular golfer had expected relative scores of -1.00 and -0.99 respectively. This means that in the 22nd most favorable tournament for the golfer he expects to take one fewer stroke than the average for that tournament. For the 23rd ranked tournament the mathematical expectation is 0.99 strokes better. A ranking is a ranking but this small difference should hardly matter. Instead of being forced to use the full integer difference (from 22 to 23) the regression analysis will use the actual

[2]The workings of logistic regression models need not concern the uninitiated reader. A special regression model is required because in the data, the left hand side variable will be dichotomous, that is, it will equal either zero (if the golfer did not enter) or one (if the golfer did enter). We will be interested in the direction of the effect, that is, does a higher expected score make it more likely to enter or less likely to enter, and the statistical significance of the effect.

expectations (-1.00 and -0.99) in calculating the effect. This should make the regression analysis more powerful.

The second difference, however, should make the combinatorial analysis more powerful. Namely, regression analysis imposes a specific algebraic structure on the relationship between the expected relative scores and the probability that a tournament is entered that is absent in the combinatorial analysis. Essentially, the combinatorial analysis is asking a question about the whole subset of tournaments that is entered while the regression analysis is asking a question about one parameter in a specific systematic relationship between expected performance and the entry decision for each individual tournament. As the results show, these effects seem to be of little consequence and this is taken by me to be a sign of the robustness of the results.

The results of these 100 regressions in Eq. (10.1) generally support the conclusions reached in the bottom panel of Fig. 9.4 in the previous chapter. As in that chapter, it is too cumbersome to list the results of all 100 regressions, but to give a sense of the results consider the ranges for the following specific statistics. The number of observations ranged from 40 to 43. Remember that of 46 tournaments, three overlap so the maximum possible is 43. Those golfers with fewer observations were not eligible for some tournaments that had restricted fields. The Cox and Snell R^2 ranged from 0 to 0.54 with an average of 0.09. The coefficient estimates, 73 of which had the expected negative sign, ranged from -12.2 to 2.6 with an average of -1.6. Our main concern is with the significance levels of these coefficient estimates.

For the closest comparison to the combinatorial results we want the probability in the left hand tail of the theoretical distribution of coefficient estimates. That is, we want the probability, called the p-value, that the coefficient estimate would be at least as low as calculated simply due to random variation if the true coefficient were actually zero. A plot of these p-values appears in the top panel of Fig. 10.1. As the figure shows, 44 of the golfers have p-values less than 0.05 (and 28 of these are less than 0.01). These numbers compare to 45 less than 0.05 and 37 less than 0.01 from the bottom panel of Fig. 9.4. The overlap between these two sets of golfers is almost exact. Only one of the 44 who are significant ($p<0.05$) in the regression analysis is not significant in the combinatorial analysis, and only

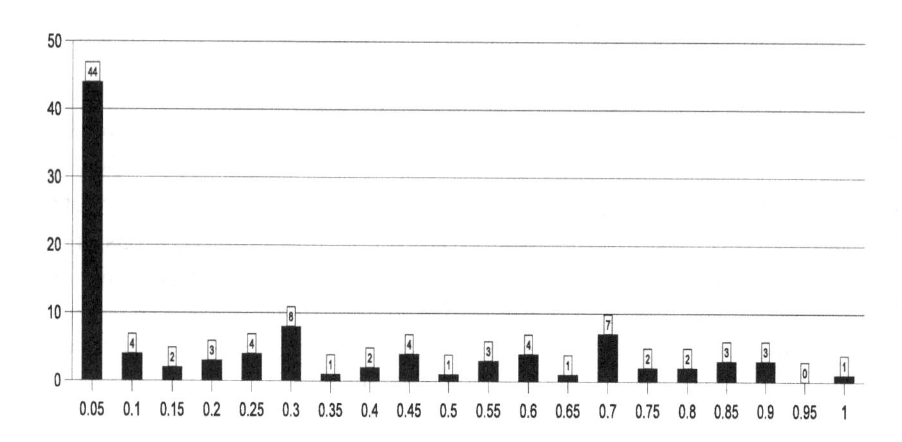

Distributions of p-values from regression estimates of Eq. (10.1)

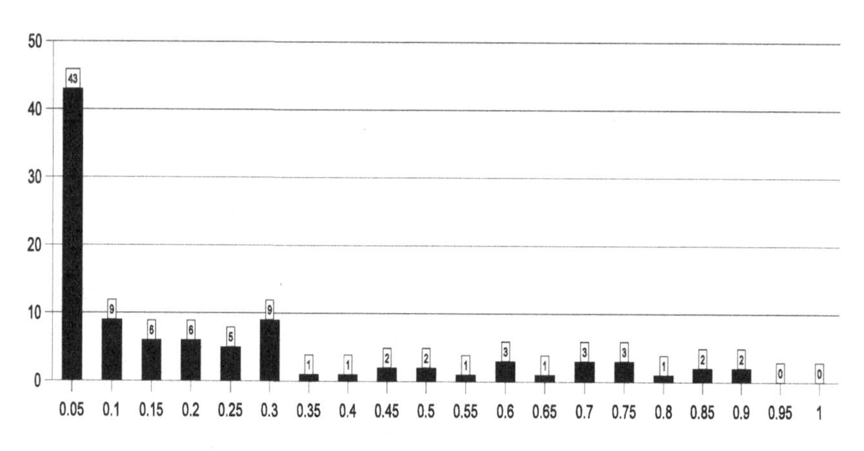

Distributions of p-values from regression estimates of Eq. (10.2)

Notes: The bar at 0.05 indicates all those whose coefficient estimates have p-values less than 0.05 ($p < 0.05$), the bar at 0.1 indicates all those whose coefficient estimates have p-values from 0.05 up to but not including 0.1 ($0.05 \leq p < 0.1$), and so on.

Fig. 10.1 Distributions of p-values.

2 of the 45 who are significant ($p<0.05$) in the combinatorial analysis are not also significant in the regression. As such, the combinatorial analysis appears to have a slightly stronger ability to capture the underlying systematic tendency hypothesized in this research, but the difference is not significant.

But regression analysis has another powerful advantage, namely, the ability to simultaneously consider more than one independent variable in a multiple regression setting. The combinatorial analysis indicated that expected relative score mattered, that the purse mattered, but that the expected earnings (which captures the strength of the rest of the field) did not matter. But these results were arrived at separately by considering only one effect at a time. So now consider Eq. (10.2),

$$\text{Probability of entering a tournament} = f(\text{constant}, \text{expected relative score, purse, expected earnings}, X)\,, \quad (10.2)$$

where the three effects examined separately in Chapter 9 can be simultaneously considered. Furthermore, the vector of control variables, X, can include control variables for the number of tournaments entered, experience, age, whether a tournament is a major, and most importantly, certain dynamic effects such as injuries, "hot hands," or the desire to cross certain relative earnings thresholds to keep playing privileges or gain entry into special tournaments.

The following dynamic effects are captured with dummy variables (or in the first case by a restriction on the sample) capturing the following considerations. First, in some cases, the known injury status of a golfer precludes entry into a tournament that may otherwise suit that golfer's skill set. In a sense, the golfer's decision rule is hierarchical, that is, if injured, do not enter, if not injured, then consider all the other aspects and decide whether to enter. As such, injured golfers are using a different decision rule and will not be modeled by Eq. (10.2). Observations on these golfer-tournament combinations are omitted from the regression reducing the usable sample from 4200 to 4099 observations.

Second, golfers may alter a planned schedule of tournament entry due to recent success (exploit a hot hand by playing, or take off to celebrate a big win) or failure (take off to work on skills, or continue playing to try to make up for lost earnings). As indicated, these effects could go either way. A set of dummy variables will try to capture the direction and significance of these considerations. A dummy variable, BIGMONEY, takes the value of one if the golfer won $100,000 or more in the previous week's tournament. A dummy variable, MISSEDCUT, takes the value of one if the golfer entered but did not make the cut in the previous week. A golfer earns zero in this case. A golfer also earns zero if he did not enter the previous tournament; this case is captured by the dummy variable, REST. The omitted category in this listing is when the golfer entered, made the cut, and earned less than $100,000 the previous week.

Arguably, the most important dynamic factor to consider is the golfer's desire to finish the year in the top 125 money winners in order to maintain exempt status for the next year. In 2006, number 125 on the money list earned $660,898. Golfers do not know this cutoff exactly but can form a pretty close approximation of how much they need to earn on average each week to meet the cutoff. Certainly, a golfer winning a tournament early in the year, and earning close to or over $1,000,000 by doing so, does not have to worry about maintaining exempt status. For others, however, it may be a year-long struggle to cross the threshold. A dummy variable, BEHINDPACE, gets the value of one for all observations in which the golfer is behind the weekly pace to make the top 125.[3]

Finally, as golfers approach the end of the year, it becomes clearer to them whether they have a chance to cross over the threshold into the top 125 (to maintain exempt status) or into the top 30 (to achieve a special status), or whether they are in danger of having the threshold move above them because of good performances by those currently below them in the ranking. For the last 5 tournaments of the year the golfers who are below either threshold but can reasonably jump above it, and golfers who are just above the threshold and in danger of falling below it are given a value of one in the dummy variable, THRESHOLD.

Equation (10.2) was estimated with a binomial logistic regression and the results appear in Table 10.1. The equation was estimated first on the

[3]However, a golfer will not be considered "behind pace" until after he enters his first tournament of the year. Many golfers choose to skip the trip to Hawaii for the early tournaments.

pooled sample with the added constraint that the effect of the expected relative score was the same for all golfers. This result is in the left column of the table. In the right column, each golfer is allowed his own coefficient of the expected score variable.

The results confirm the inferences derived from the combinatorial analysis. Consider first the left column. The higher the golfer's expected relative score, the lower the probability that he enters the tournament. The higher the purse, the greater the probability that a golfer enters the tournament. And, paradoxically, the lower the expected prize, the higher the entry probability. Golfers enter tournaments where the skills match is favorable, where the purse is high, and where, due to the presence of stiff competition, the expected prize is low. These are the same three results from the combinatorial analysis.

Several of the control variables are statistically significant in the theoretically predicted direction. With respect to the recent performance of the golfer, there is evidence that golfers may attempt to exploit a hot hand by entering a tournament after winning a large prize in the immediately preceding tournament. Having missed the cut in the previous tournament or having not played do not seem to effect the entry decision going forward.

There is also strong evidence that golfers are motivated to keep their playing privileges for the next season. If they fall behind the pace they need to earn enough to reach the top 125 on the earnings list, there is a higher probability that they enter the next tournament. Also, at the end of the season, if they are near the important thresholds of 30th or 125th on the money list, they tend to enter tournaments more than they otherwise would. The only other control variable that is significant is the number of tournaments entered by a golfer. Controls for age, experience, last year's earnings, and whether a tournament is a major were not significant.

The pooled regression in the left hand column of Table 10.1 could leave the impression that all 100 golfers in the sample can systematically choose with respect to the skills match. However, the right hand column reports results with a different coefficient on the expected score variable for each golfer. As the table shows, the significant effects from the left column are still significant. Last year's earnings and missing the cut in the previous tournament are now also significant. No other inferences are

Table 10.1 Estimates (standard errors) of Eq. (10.2).

Dependent Variable: binary, tournament entered = 1		
Equation Method	10.2 Logit	10.2 Logit
Constant	-4.484*** (0.461)	-2.806*** (1.005)
Expected relative score	-0.321*** (0.092)	individual coefficients by golfer See Fig. 10.1 bottom panel
Purse/1,000,000	0.490*** (0.044)	0.466*** (0.049)
Expected earnings/1,000	-0.001*** (0.000)	-0.002*** (0.000)
# of tournaments entered	0.111*** (0.007)	0.044*** (0.017)
career tournaments entered	-0.000 (0.000)	-0.002 (0.001)
age	-0.002 (0.009)	0.011 (0.028)
2005 earnings/1,000,000	0.002 (0.003)	0.008* (0.005)
major	0.029 (0.149)	-0.100 (0.152)
BIGMONEY	0.305** (0.134)	0.374*** (0.139)
MISSEDCUT	-0.149 (0.107)	-0.237** (0.110)
REST	-0.041 (0.085)	-0.034 (0.087)
BEHINDPACE	0.777*** (0.097)	0.775*** (0.121)
THRESHOLD	1.202*** (0.246)	1.348*** (0.256)
Cox and Snell R^2	0.152	0.178
n	4099	4099

Note: *, **, and *** indicate significance at the 10%, 5%, and 1% levels.

changed. Meanwhile the 100 individual coefficients confirm the result that roughly half of the golfers choose systematically with respect to the skills match, while the others do not. As before, the bar chart plot of the 100 p-values, this time pictured in the bottom panel of Fig. 10.1, bears a striking resemblance to those already discussed. Eighty-four of the 100 golfers have the correct negative sign and 52 of them are significant at the 10% level. Clearly, more than half of the golfers are more likely to enter a tournament that favors their individual skills, controlling for purse size, strength of competition, dynamic effects, and the other variables included in Eq. (10.2).

So, which golfers are which? Table 10.2 lists the 100 golfers whose statistics were tracked along with the associated p-value from the regression in the right hand column of Table 10.1. To interpret Table 10.2 consider the entry for Fred Couples who has a p-value of 0.050 or five percent. Fred Couples had a significant negative coefficient in the regression of Eq. (10.2) meaning that as his expected relative score in a particular tournament went up, the probability that he would enter that tournament went down. In statistics, there is always the chance that such a result could have come about because of random variation even if there were no true effect, but the p-value indicates that the chances are low (only five percent) that this would have happened in Fred Couples's case. And Couples is not an unusual case. Over 40 percent of the sample have even stronger results that his.

Chapter 9 analyzed the list statistically with the goal of determining what could explain why some golfers seem to choose tournaments wisely and others do not. Except for one variable nothing of statistical significance jumped out. Age, experience, earnings, and earnings per tournament all showed no difference with respect to whether tournaments entered was systematically or randomly related to the expected score based on the skills match.

The one variable that was significant was the number of tournaments entered, and there may be straightforward explanations for that. Golfers who do not enter the tournaments most suited to their style of play and skills may actually have to enter more tournaments to stay in the top 125 to maintain their exempt status or to get into the top 30 to receive entry into special tournaments. Furthermore, these golfers may be forced to

Table 10.2 p-values of individual golfers.

Golfer	p-value	Golfer	p-value	Golfer	p-value
Mark Hensby	0	Chad Campbell	0.029	Billy Andrade	0.219
Darren Clarke	0	Aaron Baddeley	0.032	Shigeki Maruyama	0.238
Retief Goosen	0.0002	Robert Allenby	0.032	Brad Faxon	0.256
Ian Poulter	0.0003	Tim Herron	0.034	Scott McCarron	0.261
Rod Pampling	0.0004	Stewart Cink	0.036	Justin Rose	0.268
Ernie Els	0.0004	Sean O'Hair	0.041	Ted Purdy	0.269
Padraig Harrington	0.0005	Tag Ridings	0.043	Olin Browne	0.269
Geoff Ogilvy	0.0005	J. J. Henry	0.049	Ryan Palmer	0.272
Tim Clark	0.0006	Bart Bryant	0.050	Richard S. Johnson	0.272
John Daly	0.0008	Fred Couples	0.050	Kevin Na	0.284
Jose Maria Olazabel	0.001	Bob Tway	0.051	Harrison Frazar	0.287
Luke Donald	0.001	Davis Love III	0.057	Heath Slocum	0.328
Stuart Appleby	0.002	Wes Short Jr.	0.063	Brett Quigley	0.363
Sergio Garcia	0.002	Greg Owen	0.066	Jonathan Kaye	0.413
Peter Lonard	0.003	Chris Dimarco	0.067	Kenny Perry	0.414
Vaughn Taylor	0.003	Justin Leonard	0.074	Woody Austin	0.467
James Driscoll	0.003	Lucas Glover	0.078	Arjun Atwal	0.469
Pat Perez	0.003	Scott Verplank	0.090	Daniel Chopra	0.526
Phil Mickelson	0.004	Charles Howell III	0.131	Tom Pernice Jr.	0.554
Arron Oberholser	0.005	Tiger Woods	0.134	Fred Funk	0.560
Carl Pettersson	0.005	Mark Calcavecchia	0.135	Steve Flesch	0.565
Stephen Ames	0.006	Jim Furyk	0.135	Dudley Hart	0.609
Ben Crane	0.006	Robert Gamez	0.140	Steve Lowery	0.652
Fredrik Jacobson	0.009	Brandt Jobe	0.143	Bob Estes	0.692
Rory Sabbatini	0.010	Jason Bohn	0.160	J. L. Lewis	0.693
Tim Petrovic	0.012	Jason Gore	0.167	Joe Durant	0.709
David Toms	0.014	Jeff Brehaut	0.175	Billy Mayfair	0.734
Bernhard Langer	0.015	Carlos Franco	0.188	Joey Snyder III	0.737
Vijay Singh	0.015	Brian Davis	0.194	Jerry Kelly	0.755
Mike Weir	0.015	Joe Ogilvie	0.198	Jeff Sluman	0.820
Tom Lehman	0.016	Bo Van Pelt	0.204	Loren Roberts	0.846
Steve Elkington	0.016	John Rollins	0.211	John Senden	0.857
Adam Scott	0.020	Charles Warren	0.217	Joey Sindelar	0.864
K. J. Choi	0.023				

Note: The p-value reported is the probability that the individual golfer's coefficient from Eq. (10.2) would be at least as low as calculated due to random variation if the true coefficient were zero.

enter tournaments that come late in the calendar year even though they know that the match between the course and their own skills is not optimal. Simply put, certain golfers may be, perhaps unexpectedly, playing more year-end tournaments that do not match their skill sets, thus leading to the correlation between the number of tournaments entered and the inability to choose tournaments favorable to one's skill set.

This chapter and the preceding one examine the choice made by professional golfers of which tournaments to enter. By looking at tournament level data for the top 100 money winners of 2005 during the 2006 season, the specific strengths of the individual golfers and the specific required strengths at any one tournament site can be determined. This information allows one to calculate the expected relative performance of each golfer at each tournament. If golfers ignore this information and choose which tournaments to enter based on other factors and, therefore, randomly with respect to these calculations, the null hypothesis of random tournament selection will not be rejected.

Two methods were used to examine the entry choice. In Chapter 9, a combinatorial analysis compared the subset of tournaments chosen to the random distribution of possible subsets. In Chapter 10, a regression analysis examined whether a systematic relationship between expected relative performance and the entry decision could be uncovered.

Forty-six tournaments were examined, individual golfers choosing to play in as few as four and as many as 32 of them. Based on the combinatorial analysis, the results pictured in Chapter 9 show that close to half the golfers do systematically choose which tournaments to compete in based on their skill strengths and the requirements of the specific tournament. Based on the regression analysis in Chapter 10, the results pictured in Figure 10.1 and listed in Table 10.1 confirm and extend those of the combinatorial analysis. This matching of talents and needs is an important issue in labor economics. The result also suggests the next step in this research agenda, namely, to discover the characteristics that allow certain golfers to fall into the group that chooses systematically while others seemingly choose randomly. A quick check of some potential factors such as age, earnings, experience, and number of tournaments entered, did not reveal significant differences except for the number of tournaments chosen. Further analysis of this question is beyond the scope of this book and is left as a suggestion for future research.

Chapter 11

To Bet or not to Bet: Sports Gambling, Golf, and Efficient Markets[1]

Economists examine gambling markets trying to uncover evidence of profitable opportunities, or if failing to do so, providing evidence of a type of market efficiency.[2] This chapter will flesh out the arguments that economists pursue and look at new evidence encompassed in a casino's betting line for posted odds on PGA TOUR events. The chapter proceeds by first discussing two separate dimensions of financial market efficiency, namely informational efficiency and transactional efficiency. Following this, the analogy between financial markets and betting markets is explained along with the institutions in sports gambling. The more familiar institutions in point spread betting for basketball and football, and parimutuel betting at horse races are described first for background and for comparison sake. Posted odds betting on golf is slightly different but the same underlying arguments about market efficiency apply. Finally, the evidence from the 2002 professional golf season is presented to assess the two types of efficiency that are introduced.

[1]This chapter draws from research originally published in Shmanske, S. (2005). Odds-Setting Efficiency in Gambling Markets: Evidence from the PGA TOUR, *Journal of Economics and Finance*, 29(3), pp. 391-402, and presented in a workshop at California State University, East Bay and at the Western Economic Association International meetings in Vancouver, July 2004.

[2]The literature is vast. An excellent summary and critique of the methods and results is in Sauer, R. D. (1998). The Economics of Wagering Markets, *Journal of Economic Literature*. 36(4), pp. 2021-2064.

11.1 Two Types of Market Efficiency

The first type of market efficiency concerns the ability of the market price to capture and relay information about the quality or value of any financial asset. Recognition of this property of market prices goes back at least to Armen Alchian [1974] who proposed that the current market price must be an unbiased predictor of the future price or else positive profit opportunities would exist.[3] The argument is fairly simple. Suppose that the current market price was systematically biased to be below the average future price. If so, it would be profitable to buy the asset, hold it, and sell it in the future at a price that was, on average, above the purchase price. Of course, any price increase would also have to cover the foregone interest, any holding costs, and any transaction costs including commissions and the like. These items are actually only technicalities that may obscure but do not undo the basic principle involved. The basic principle is that any bias is not expected to persist because arbitragers will attempt to buy such underpriced assets and their purchases will drive up the current price until the discrepancy no longer exists. In this way the current price signals in an unbiased manner all the publicly available information about the asset.

When new information about an asset comes out, market participants will change their expectations of the future price and, therefore, the pre-existing current price will be a biased predictor until those profit seeking market participants buy or sell the asset thus bringing its price back into line with the future expected price. When economists or finance specialists speak of efficient markets or the efficient market hypothesis they are usually referring to the ability of the price to fully incorporate all publicly available information as quickly as possible. When economists started to test these propositions in the 1970's, the best they could do was to use daily closing stock price quotes. It appeared that new information was incorporated in the stock's price within one day. Later, when computers made tracking the actual minute to minute transactional price data possible, it was discovered that new information was incorporated into the price in less than one hour and possibly as fast as 15 minutes or so. This, of course, means that by the time you hear news on the radio or read it in the

[3]See, Alchian, A. (1974). Information, Martingales and Prices, *Swedish Journal of Economics*, 76(1), pp. 3-11.

newspaper or even online it has already been incorporated into the market price. Unless you are the first to find out the information or are acting on private information that is not available to the rest of the market you should probably avoid trying to arbitrage stock prices. You won't be able to systematically buy low and sell high, you will probably end up buying at the average and selling at the average and paying commissions to the brokerage for the privilege.

The informational efficiency in market prices aside, there is another aspect of efficiency which can be called transactional efficiency. Simply put, how much will it cost in commissions and fees to enter (and exit) the stock market, or any market for that matter. Harold Demsetz [1968] studied this aspect of transactions and suggested looking at the bid-ask spread in pricing as a measure of both transactions costs to the buyer and profitability to the broker.[4] Bid-ask spreads are competitively set but must cover the cost (including the risk of price changes) to the broker. If the market has frequent trading, the broker does not have to hold (or be short) any securities for too long a time (the longer the time held the more risk that the price will change). Alternatively, in a thin market, that is, one with less frequent trading, the broker may have to hold securities for longer periods and thus be exposed to more risk that new information will cause the price to move outside the established bid-ask spread causing losses to the broker. The bottom line in this research is the intuitively appealing result that bid-ask spreads shrink when the market volume grows. Thus, thin markets are less efficient in the sense of higher bid-ask spreads and a larger implicit profit margin to the broker. Wagering on golf should be less efficient in this sense than wagering on basketball or football because, for golf, the amount of money wagered is much smaller. This much is obvious, but within golf itself a more subtle natural experiment exists because the overwhelming popularity of Tiger Woods spills over into gambling markets. When Tiger Woods is scheduled to compete, the betting volume goes up, and the implied efficiency of the odds prices in the transactions costs sense (and the predictive sense) can be compared to tournaments in which he is not competing.

[4]See, Demsetz, H. (1968). The Cost of Transacting, *Quarterly Journal of Economics*, 82, pp. 33-53.

11.2 From Financial Market Efficiency to Sports Gambling

The analogy from financial markets to gambling markets follows from a comparison of a gambling market equilibrium with the absence of systematic profitable opportunities to an asset market equilibrium in which the current price of an asset must be an unbiased predictor of the future price to prevent arbitragers from profiting. It behooves us to first outline how a typical sports betting regimen works. Consider betting on a basketball game. The usual method is that the bettor must give the casino 11-10 odds. This means that the bettor will put up $11 and the casino will put up $10, with the casino acting as the escrow agent. If the casino wins the bet, they just keep the bettor's $11. If the bettor wins, the receipt for the bet is worth $21, representing a gain of $10 because $11 of the $21 is the bettor's original gamble. The casino can act as a pure broker and not take on any risk if it can achieve a "balanced book" in which there are the same number of bets on the favorite as there are on the underdog. Supposing there is one bet on each, then the casino collects $22 and will have to pay only $21 to the winner, thus collecting $1 with no risk. Meanwhile, the bettor will have to win 11 out of 21 bets (or 52.4%) to break even.

Ordinarily the casino would not achieve a balanced book because more people would bet on the favorite, and if the favorite won more often as would be expected, the casino would lose more bets than it would win, and bettors would win as long as the favorite won at least 52.4% of the time. So casinos develop a system of point-spread gambling in which the favorite has to win by more than the point spread for a bet on the favorite to win. This system makes it harder for the betting public to predict the winning side of the bet and the money bet on each side of the bet will even out. Proper establishment of the point spread will make the bet a toss up that the bettor will win about 50% of the time.

Suppose that the point spread was too low as was the case above before a point spread was even considered. With no point spread, or what is the same thing, a point spread of zero, or any time the point spread is too low, there will be too many bets on the favorite from the casino's point of view. But the point spread can be adjusted. As the point spread increases, fewer bettors will think it likely that the favorite will win by a large enough amount to cover the spread and more bettors will think it likely that the underdog can get close enough, or win outright, and the book will be brought into balance. If the point spread is too high the adjustment works in the same manner in the opposite direction. The equilibrium of this process is when the point spread is equal to the winning margin that is expected by the betting public, or to borrow some language from the financial markets, when the point spread is an unbiased estimate of the game outcome.

If the point spread were systematically biased up or down, then in straightforward analogy to the arbitragers in the case of too high or too low a current asset price, profits would be available to savvy bettors who could identify the bias, and the casinos would stand to lose too many bets. There is a slight complication in the analogy. In financial asset markets, the quality of the asset stays the same and the price adjusts. In point spread gambling markets the price stays the same (risk $11 to win $10) but the "quality" of the bet adjusts depending upon how many points the bettor on the favorite has to give to the underdog. But the complication should not confuse us. It is really the quality/price ratio that adjusts to bring the market into an equilibrium with no systematic profit opportunities. In financial markets the price adjusts, in betting markets the point spread, or quality, adjusts.

To summarize the analogy with respect to informational efficiency one would say that the point spread is an unbiased predictor of the final game score differential in the same way that an asset's current price is an unbiased predictor of the future price of the asset. What about the transactional efficiency?

Demsetz's argument about transaction costs implies looking at the costs of setting up a zero risk portfolio. In the case of a share of stock, if an investor simultaneously bought and sold the asset, he would end up bearing no risk of price movement of the asset. Of course the investor would have to pay the asking price and would only receive the bid price and the cost would be equivalent to the bid-ask spread on the stock. In point spread gambling it is possible to set up a "zero risk" portfolio by taking both sides of the bet and assuring that one bet will win and one will lose.[5] The bettor would pay $22 and receive back $21, with the one dollar difference being analogous to the lost bid-ask spread from the financial transaction. And just as it would seem pointless to simultaneously buy and sell the same asset it would seem just as pointless to take both sides of the point spread gamble. The one dollar loss out of $22 risked represents a transaction cost of about 4.5%. This is probably a higher percentage than represented in a typical bid-ask spread, but as we shall see, it is low compared to other forms of sports gambling, and is largely responsible for the popularity of this type of gambling on sports.

By way of contrast, in the typical system of betting on horse races at racetracks, called parimutuel, the transaction cost is on the order of 15 or 20 percent. Consider bets on which horse will win a race. The house, the racetrack in this case, collects all the bets on the different horses to win, takes out its percentage (called the take or vig) and pays out the rest to the bettors on the winning horse. The house is not at risk in this system and is really only acting as an escrow agent for the bettors who are actually betting against each other. It should not be surprising that the transaction cost to the bettors at the track is higher than for point spread gambling at a casino. The track's take has to pay for the costs of staging the races themselves whereas the casinos do not stage the events that they accept bets on. Along the same line of argument, off-site gambling or internet gambling on horse racing typically has a much lower vig to the house because of the low overhead cost. As we shall see in the next section the transaction costs for betting on PGA TOUR golf are much higher than for either point spread betting or parimutuel betting at the race track.

11.3 Transactional Efficiency in Wagering on PGA TOUR Golf

Wagering on PGA TOUR golf works differently than the wagering in point spread gambling or in parimutuel horse race gambling. It consists of the weekly posting of odds by casinos to win that week's tournament for a variety of golfers and for the "field," which includes everyone else in the

[5]Tied bets are a possibility that adds a minor technical correction to the argument. This has been explored in Shmanske, S. (1991). Tied Bets, Half Points, and Price Discrimination, *Kentucky Journal of Economics and Business*, 11, pp.43-54.

tournament without specifically listed odds. Bettors can place bets from the time of completion of the previous tournament up until the Thursday morning start of the current tournament. The best way to get a feel for the method is to consider a typical odds sheet. Table 11.1 shows how the odds for a typical tournament are listed using the example of the 2002 Advil Western Open.[6] This odds sheet is particularly interesting because it shows the odds with and without Tiger Woods due to his late withdrawal from the tournament. For most of the other tournaments during 2002, the odds as originally posted did not change over the course of the week. Table 11.1 illustrates a tendency that was consistent throughout the 2002 season. The worst odds are always offered on Tiger Woods when he is competing, and if he is not competing, the worst odds are always offered on the field.

Golf wagering differs from the wagering described in the previous section. In point spread gambling and parimutuel betting the house is acting as more of a broker than a bettor, in that it is paying off the winning bets with the proceeds of the losing bets and incurring no risk. But by taking bets at posted odds, the house is actually betting against the customer. Although the house may endeavor to match the proportion bet on each golfer to the odds offered, there is generally no guarantee that the money collected on losing bets will cover the required payout to the winners. Compared to parimutuel betting, the house is at greater risk, and we would expect the transaction costs to the bettor to increase to compensate for the extra risk.

The greater risk to casinos from posted odds bets, and perhaps also from the comparative thinness of the market, shows up in a higher "price" of placing a bet. At the track, for example, a portfolio of bets could be made that would assure that the bettor would win one of them, but if the track takes 20% of the pool before proportionately dividing the rest among the winners, it would cost \$1.25 to return \$1.00. For point spread gambling, the portfolio includes bets on each of the favorite and the underdog guaranteeing a winning bet (except in the case of a tie), but it would cost \$1.0476 (= 22/21) to return \$1.00. For the posted odds on PGA TOUR golf tournaments, one could also form such a portfolio, but for the

[6]All the odds used as data were graciously supplied by Pete Korner from the archives of Las Vegas Sports Consultants.

Table 11.1 Odds for 2002 Advil Western Open.

Cog Hill Golf and Country Club July 4-July 7, 2002
(TakeDown Date: July 4, 2002 4:00 AM)

	Open (July 1)	Current (July 4)	note
Tiger Woods	6/5	XXX	Withdrew 7/2
Vijay Singh	12/1	10/1	
Justin Leonard	18/1	12/1	
Davis Love III	20/1	15/1	
David Toms	20/1	15/1	
Charles Howell III	25/1	20/1	
Mike Weir	25/1	20/1	
Scott Hoch	25/1	20/1	
Stuart Appleby	30/1	25/1	
Nick Price	30/1	25/1	
Scott Verplank	30/1	25/1	
Robert Allenby	30/1	25/1	
K. J. Choi	35/1	30/1	
Kenny Perry	35/1	30/1	
Loren Roberts	35/1	30/1	
Jerry Kelly	35/1	30/1	###winner###
Rocco Mediate	35/1	30/1	
Bob Estes	35/1	30/1	
Cameron Beckman	40/1	35/1	
Steve Flesch	40/1	35/1	
Peter Lonard	45/1	40/1	
Scott McCarron	45/1	40/1	
Lee Janzen	45/1	40/1	
Steve Stricker	45/1	40/1	
Ian Leggatt	50/1	45/1	
Frank Lickliter II	50/1	45/1	
Steve Elkington	50/1	45/1	
Jeff Sluman	50/1	45/1	
Chris Riley	60/1	50/1	
Skip Kendall	60/1	50/1	
Field	3/1	even	

Source: Las Vegas Sports Consultants.

final odds as posted for the tournament in Table 11.1, the cost would be a whopping $1.56 to return $1.00.

To see how this calculation is made consider a small numbers example in which there are four contestants upon which to bet at the posted odds given in the Odds column of Table 11.2. If a bettor placed a one dollar bet

Table 11.2 Posted odds betting with four contestants.

Contestant	Odds	Better Odds
A	1-1 (even)	1-1 (even)
B	2-1	3-1
C	3-1	4-1
D	4-1	5-1

on each of the contestants, then he would be assured of winning one of them but his payoff would be different depending upon who won. To take away all risk we want a portfolio of bets that returns the same amount of money regardless who wins. In this case, using 60 as the least common denominator the bettor could bet $30 on A, $20 on B, $15 on C, and $12 on D and no matter who won would receive back $60 for the winning bet stub. If A won paying even money, the bettor would get back his $30 bet on A along with $30 of winnings. If B won, the bettor would get back his $20 bet on B along with $40 of winnings since the odds on B were 2-to-1. If C won, the bettor would get $15 plus $45, and if D won the bettor would get $12 plus $48. The cost to this bettor of getting back $60 is the total amount he wagered, in this case, (30 + 20 + 15 + 12 = 77) $77. Although fractional amounts bet are not typically allowed, dividing by 60 we find that it would cost (77/60 = 1.2833) slightly over $1.28 to win $1.

If there was another casino competing for this business they might do so by offering better odds as listed in the right column in Table 11.2. By making the payoff on bets on B, C, and D better, they are reducing the cost of the set winning portfolio. As quick calculation shows, the gambler could now wager $30 on A, $15 on B, $12 on C, and $10 on D and collect $60 no matter who won. The total amount wagered is now $67 to get back $60 for a cost of (67/60 = 1.1167) a little under $1.12 to win $1.

This simple example shows how increases in the odds offer the betting public a better deal. Now, going back to the actual odds that were offered by the casinos, the implied price of a guaranteed return of $1.00 can be calculated for each tournament, and for subsets of tournaments. In 2002, these prices ranged from a low of $1.36 to a high of $1.81. Over all 36 tournaments in the sample, the average price is $1.51. Generally speaking, betting on golf is not nearly as advantageous to the bettor as betting on football, basketball, or horse racing. Neither is it as consistent in terms of the house's take. Do not forget, however, that the house has to cover its risk that the losing bets might not yield enough funds to pay the posted odds to the winning bettors.

We can now move to our hypothesis test concerning transactional efficiency in golf gambling markets. If greater amounts are wagered on a tournament, it might allow the casino to offer more favorable odds, much as transactions costs, measured by bid-ask spreads, are lowered in thick markets. Tiger Woods competed in 13 of the 36 tournaments in the sample, and heightened the interest and betting volume of golf fans for those tournaments. If the implied price of a portfolio of bets guaranteeing a return of $1.00 is lower in those tournaments, we have evidence consistent with the proposition that thick markets are more efficient than thin markets in the sense of lower transactions costs. The evidence, however, actually goes against this proposition. In the 13 tournaments containing Tiger Woods, the implied price of a portfolio of bets that would return $1.00 is $1.62, while in the other 23 tournaments in the sample the implied price was $1.50, a difference that was not statistically significant. If anything, based on the point estimates of $1.62 and $1.50 it appears that the tournaments containing Tiger Woods are worse bets on average than the other tournaments. Perhaps the house uses the increased demand to bet on these tournaments, (or the lower elasticity of demand for those who want to bet on, or against, Tiger) as an opportunity to raise the implied price. As we shall see below, the actual outcome of bets placed on Tiger Woods is a loss to the bettors. Perhaps fans of Tiger Woods are placing bets on him for consumptive reasons even when his actual probability of winning is much less than that implicit in the odds that are offered. The 2002 season was during Tiger Woods' early period of dominance and high popularity. Interesting research could easily be done along these lines for other years in his career.

11.4 Informational Efficiency in Wagering on PGA TOUR Golf

11.4.1 *Naive betting strategies*

One of the main themes in the existing literature on sports gambling concerns whether there are identifiable biases in bettor behavior or in posted odds that can lead to positive profit strategies. If there are, then the betting market would not be in equilibrium in the efficient market sense. In particular, several analysts of parimutuel betting on horse racing have uncovered what has become known as the favorite-longshot bias.[7] Evidently, less money, proportionately, is bet on favorites at the track than their actual winning percentages would justify, while too much is bet on longshots, compared to their actual winning percentages. Whether or not a profitable betting strategy exists once the track's take is considered is not clear. What about gambling on golf?

Table 11.3 lists the actual returns and the implied prices from several betting strategies if followed over the course of the 2002 PGA TOUR season. Several strategies come to mind. One strategy is to always bet on Tiger Woods. Its complement is to always bet on everyone except Tiger. Table 11.3 shows that neither of these strategies works. Tiger competed in 13 tournaments so betting on Tiger would entail betting $13 if a minimum bet of $1.00 could be laid. Tiger actually won three of the tournaments he entered and would have returned $9.50 in total. Thus it would cost $1.37 for each dollar returned with this strategy. Meanwhile, always betting on

[7]See Asch P., Malkiel, B. G., and Quandt. R. E. (1982). Racetrack Betting and Informed Behavior, *Journal of Financial Economics*, 10, pp. 187-194, Thaler, R. H. and Ziemba, W. T. (1988). **Anomalies** Parimutuel Betting Markets: Racetracks and Lotteries, *Journal of Economic Perspectives*, 2(2), pp. 161-174, Hurley, W. and McDonough, L. (1995). A Note on the Hayek Hypothesis and the Favorite-Longshot Bias in Parimutuel Betting, *American Economic Review*, 85(4), pp. 949-955, and Sauer, R. D. (1998). The Economics of Wagering Markets, *Journal of Economic Literature*, 36(4), pp. 2021-2064.

everyone else except Tiger would mean placing 384 $1.00 bets to return a total of $167. Betting against Tiger costs $2.30 for every dollar returned.

Table 11.3 Outcomes for naive betting strategies.

Strategy	Amount Bet	Amount Returned	Cost to win $1
Always Bet on Tiger Woods	$ 13	$ 9.50	$1.37
Always Bet against Tiger Woods	$ 384	$ 167	$2.30
Always Bet on the Favorite	$ 36	$ 30.50	$1.18
Longshot Bets	$ 812	$ 247	$3.29
Shorter Odds Bets (not favorite)	$ 327	$ 228	$1.43
Always Bet on the Field	$ 34	$ 50.27	$0.68

Notes: Amounts in the table are based on minimum bets of $1.00 placed each time.
Source: Author's calculations.

Always betting on the favorite, whether or not Tiger is in the tournament does slightly better than always betting on Tiger. The favorite won six of the 36 tournaments (Phil Mickelson, at 7/1, was the longest odds favorite to win) for a total return of $30.50 on the placement of 36 $1.00 bets, that is, a cost of $1.18 to win $1.00.

Two other strategies take a little more to explain. The first attempts to capture those with long odds. Starting from the player with the highest posted odds, for example, Skip Kendall at 50/1 in Table 11.1, bet on all the long odds players, up to the point where if any one of the golfers bet upon wins, the return to the bettor would be positive. In Table 11.1 this entails betting on everyone up to and including Stuart Appleby at 25/1. This would mean betting on 22 golfers, say one dollar apiece, (although the minimum bet is probably $10) and getting at least $26 back if one of those 22 golfers wins. For this tournament the return would be positive, $31 returned for every $22 bet because Jerry Kelly won and paid 30/1. As Table 11.3 shows, however, over the whole year this is not a winning strategy. This strategy entails placing 812 $1.00 bets and getting a total return of only $247. This costs $3.29 to return $1.00.

The next strategy attempts to capture those golfers other than the favorite with the shortest odds and presumably the best chance to win. The strategy is to bet on as many of the shortest odds golfers as possible while still assuring a positive return if any of them win For example, in Table 11.1, starting with Justin Leonard at 12/1 one could place at most 12 $1.00 bets and get a return of at least $13 even if the lowest odds player won. Picking off the top 10 except for the favorite Vijay Singh, bets are placed down to Robert Allenby at 25/1. Since there is no obvious objective way to choose only two of the six golfers offered at 30/1, do not bet on any of them. This strategy, if carried out for the full year, entails placing 327 $1.00 bets and returning $228. This, too, is a losing strategy costing $1.43 for every dollar returned.

Finally, however, consider the not so exciting strategy of always betting on the field. Two tournaments of the 36 had exclusive entry conditions and offered odds on each entrant with no field bet. Thus, 34 $1.00 bets would have been made. The field won on 15 occasions, returning $50.27. This, indeed, would have been a winning strategy in 2002, costing only $0.68 for each dollar returned. This may simply be a small sample anomaly; the 2002 season set a record for the most first-time winners of PGA TOUR events. In 2002, the highest odds ever offered on the field were 8/1 and the lowest odds were 3/5. The field as a 3/5 odds-on favorite won two of the four times it was offered.

There is a correspondence between these results and the aforementioned favorite-longshot bias at the track. The calculations in Table 11.3 show that the long odds bets pay off much less frequently than the probability implied in the long odds, and the short odds bets, especially on the field, pay off more often than the probability implied in the odds. This is the same pattern that shows up at the track. Betting on the field is simply not very lucrative when compared to the potential loss. At 3/5 odds one has to risk $500 to win $300, whereas a 30/1 longshot entails only a risk of $10 to win $300 and allows the bettor to root for a player whose name is probably more familiar to the bettor.

The bottom line for this section is that with the possible exception of always betting on the field, no profitable simple betting strategies were uncovered for the 2002 season. Whether the bias existing for the field bet in 2002 would carry over to other years will remain a research project for an enterprising golf gambler to pursue. I can only urge the reader to do the research before betting the house on this strategy. For my part, the one result of a profitable strategy for 2002 is not enough to overturn my theoretical understanding of efficient markets theory and my experience with looking at the evidence with respect to sports gambling in other settings. This is not to say that I wouldn't be convinced by looking at more evidence, or that I would discount the possibility that putting in dozens or hundreds of hours of research looking at the statistics might be fruitful, it is just that I would be surprised if the simple, naive strategy of always betting the field would hold up as profitable in future years.

11.4.2 *Predicting the order of finish*

Besides arguing that no simple betting algorithm could be profitable, the efficient markets literature also suggests that the betting odds or point spread should capture beforehand all of the publicly available information about the outcome of a sporting event. This suggests examining the statistical relationship between the actual outcome and the outcome predicted in the odds. Consider the following equation:

$$\text{RANKOUTCOME} = b_0 + b_1\text{RANKINODDS} + e\,. \qquad (11.1)$$

RANKOUTCOME is the order of finish among those with posted odds. For example, a second place finish in the tournament would get a rank of 1 if the tournament winner was from the "field." RANKINODDS is the expected finish rank based on the posted odds. The regression coefficients are the b's, and e is an error term. If the odds ranking predicts perfectly, then b_0 and e would be zero, b_1 would equal one, and the R^2 would be one. The actual results are listed in Table 11.4. As you can see, there is a significant positive relationship between the actual finish order and the finish order predicted in the odds, but the adjusted R^2 of .055 shows that most of the variation in the finish order is left unexplained.

Table 11.4 Regression coefficients (t-statistics in parentheses).

Equation	(11.1)	(11.2)	(11.3)
Sample	All n=1034	All n=971	All n=971
Method	O.L.S	O.L.S.	O.L.S.
Dependent Variable	RANKOUTCOME	RANKOUTCOME	RANKINODDS
constant	10.43*** (23.59)	-14.14 (-0.49)	51.36** (1.916)
RANKINODDS	0.217*** (7.80)	0.154*** (4.42)	
DRIVDIST		0.073** (1.94)	0.022 (0.62)
DRIVACC		5.613 (1.18)	12.25*** (2.79)
GIR		-5.68 (-1.18)	-12.37 (-2.78)***
PUTTPER		9.63 (0.68)	-7.308 (-0.55)
BIRDIEAVERAGE		-0.130 (-0.20)	-1.193 (-1.99)**
PARBREAKERS		-55.61** (-1.90)	-97.51 (-3.62)***
RECENTFIRSTS		-0.156 (-0.86)	-0.23 (-1.37)
CAREERFIRSTS		-0.084 (-1.47)	0.127 (2.39)***
RECENTTOP10		0.075 (1.26)	-0.591 (-11.32)***
Adjusted R^2	.055	.067	.312

Note: *, **, and *** indicate significance at the 10%, 5%, and 1% levels.

Even looser correlations were found using a variety of different functional transformations. In particular, when the dependent variable was the rank among all in the tournament as opposed to the rank only among those with odds, or when the explanatory variable was the implied, by the odds, probability of winning instead of the implied rank finish order, the R^2 fell.

The low R^2 is not necessarily bothersome. The tests of informational efficiency do not require a perfect prediction, or even a close one. Rather, the tests require an *unbiased* prediction, one that cannot be improved in any systematic manner with information, other than that included in the odds, available at the time bets must be placed. If something is inherently very hard to predict due to many unknowable random factors, then the prediction equation will have a low R^2 but informational efficiency can still hold.

However, if information other than that in the odds adds to the explanatory power of the prediction equation, then the possibility of constructing a betting algorithm exploiting that information presents itself. For example, suppose the skill of driving distance was added to Eq. (11.1) on the right hand side. Suppose further, that it had a statistically and economically significant negative coefficient. Then, the equation would mean that controlling for the ranking that is predicted by the odds, long drivers would finish with even lower numbered rankings (that is, better, since a first place ranking of one is better than a second place ranking of two). This would suggest a betting algorithm of placing bets on long drivers whose higher than justified odds make them attractive bets. To explore this possibility Eq. (11.2) was estimated.

$$\begin{aligned}\text{RANKOUTCOME} = {} & b_0 + b_1\,\text{RANKINODDS} + b_2\,\text{DRIVDIST} + \\ & b_3\,\text{DRIVACC} + b_4\,\text{GIR} + b_5\,\text{PUTTPER} + \\ & b_6\,\text{BIRDIEAVERAGE} + b_7\,\text{PARBREAKERS} + \\ & b_8\,\text{RECENTSFIRSTS} + b_9\,\text{CAREERFIRSTS} + \\ & b_{10}\,\text{RECENTTOP10} + e\,. \qquad (11.2)\end{aligned}$$

DRIVDIST, DRIVACC, GIR, and PUTTPER measure the golfer's abilities in the skills of driving, approach shots and putting. As shown in previous chapters, these are the most important skills of golfers, and those most closely correlated with winning purses. Another possible factor in winning outright is the ability to make an exceptionally low score in order to beat out 143 other golfers. This factor is captured in BIRDIEAVERAGE and PARBREAKERS, respectively, the number of birdies per 18 holes and the percentage of holes on which the golfer scores under par. The last three variables capture the recent and career success of the golfer in question.[8] If the odds as offered by the casino neglect consideration of any of these factors, they will show up as significant regressors in Eq. (11.2).

Table 11.4 lists the results of Eq. (11.2). The sample is slightly smaller because the skills statistics were not available for all of the golfers with posted odds. The constant term is no longer significant and the RANKINODDS variable is still significant and positive. But the set of extra explanatory variables hardly makes any improvement. PARBREAKERS and DRIVINGDISTANCE are both significant, but none of the others are, and the adjusted R^2 goes up only to .067. Interestingly, the coefficient of DRIVINGDISTANCE is positive, meaning that long drivers finish worse than predicted in their odds. The betting algorithm that would be suggested is to place bets on golfers who are shorter off the tee as the odds offered on them seem to underestimate their actual performance. However, the effect is small. Since the coefficient is 0.07, it would take about 14 yards of decreased driving distance (about a 5% change based on an overall average of 280 yards) to improve one's finish rank by one place (about a 3% change if there are 30 or so golfers with posted odds). This would be an inelastic effect, and arguably not strong enough on which to base a betting algorithm.

The underwhelming improvement in the explained variation suggests that any information contained in these extra variables is already captured by the RANKINODDS variable, a result consistent with the efficient markets view. In this case, the efficient markets conclusion is even stronger than at first glance, because the skills statistics used in the regression covered the whole 2002 season, and would not have been available in the same form at the time of each competition.[9]

[8] All the golf statistics come from *2003 PGA TOUR Media Guide*.

[9] If these preliminary results had been more favorable, an additional step would have been required to develop a betting algorithm. Instead of using year long statistics, the regression would have to use weekly updated current statistics in a manner similar to that used with some success by Zuber, Gandar, and Bowers [1985] for NFL football. The year-end

It is not the case that golfer's skills and past performances are irrelevant to the outcome. The odds offered by the casino must have some relationship to skills and past performances. Consider Eq. (11.3):

$$\begin{aligned} \text{RANKINODDS} = \; & b_0 + b_1 \text{ DRIVDIST} + b_2 \text{ DRIVACC} + b_3 \text{ GIR} + \\ & b_4 \text{ PUTTPER} + b_5 \text{ BIRDIEAVERAGE} + \\ & b_6 \text{ PARBREAKERS} + b_7 \text{ RECENTSFIRSTS} + \\ & b_8 \text{ CAREERFIRSTS} + b_9 \text{ RECENTTOP10} + e \, . \end{aligned} \tag{11.3}$$

The results of Eq. (11.3) are in the third column of Table 11.4. Eq. (11.3) shows that the golfer's skills and past winning success are correlated to the odds offered, with an adjusted R^2 of .312. Surprising is the fact that putting skill seems statistically unimportant in setting the odds. Meanwhile, recent top-ten finishes seem statistically to be very important. Evidently, the skills and recent performances are not irrelevant to the setting of odds, but as Eq. (11.2) shows, once the RANKINODDS variable as set by the odds is included, the skills add no new information.

The final test for this section comes from separating the sample based upon whether Tiger Woods is competing. Table 11.5 lists the results of estimating Eq. (11.2) in the two split samples. More of the variation is explained when Tiger Woods is in the sample, but the results are not overwhelming. In terms of the individual variables, the RANKINODDS variable maintains its significance in both samples, and that is the only consistency across the samples. DRIVDIST is significant in the sample without Tiger, DRIVACC is significant in the sample with Tiger, and so on. There does not seem to be any "Tiger Woods effect" with respect to biases in the information imbedded in the odds. There may be a slight effect with respect to the amount of information in the odds as measured by the adjusted R^2. This is probably due to the fact that Tiger Woods was always ranked number one if he was competing, actually won three out of 13 times, and was among the leaders in most of the other cases. It may also be the case that the higher betting volume when Tiger is in the tournament enables (or forces?) the odds makers to do a better job.

statistics contain more information than the weekly updates, but since the year-end statistics did not add to the predictive power, experiments with the weekly updates were not attempted.

Table 11.5 Regression coefficients (t-statistics in parentheses).

Equation	(11.2)	(11.2)
Sample	Tiger in n=395	Tiger out n=573
Method	O.L.S	O.L.S.
Dependent Variable	RANKOUTCOME	RANKOUTCOME
constant	-103.0**	27.75
	(-2.08)	(0.77)
RANKINODDS	0.163***	0.118***
	(2.72)	(2.69)
DRIVDIST	0.036	0.111
	(0.56)	(2.34)
DRIVACC	16.72**	2.215
	(2.10)	(0.37)
GIR	-16.97	-2.209
	(-2.10)**	(-0.36)
PUTTPER	57.76	-14.36
	(2.35)**	(-0.82)
BIRDIEAVERAGE	-0.44	-0.043
	(-0.39)	(-0.06)
PARBREAKERS	25.28	-106.40***
	(0.52)	(-2.87)
RECENTFIRSTS	-0.27	-0.011
	(-1.09)	(-0.03)
CAREERFIRSTS	-0.109	-0.039
	(-1.23)	(-0.50)
RECENTTOP10	0.028	0.065
	(0.30)	(0.76)
Adjusted R^2	.096	.049

Note: *, **, and *** indicate significance at the 10%, 5%, and 1% levels.

11.4.3 *Predicting the winner*

It is possible that the odds or the implied ranking by the odds does not do well in predicting the finish order of the contestants, because the bets pay off only in the case of a win, thus making the rank order from second place down irrelevant to the bettor. This suggests that the dependent variable should capture winning versus not winning as opposed to the actual ranking against all other contestants. Furthermore, the explanatory variable should not measure the implied rank order of the contestants either. Instead, it should measure the probability of actually winning the tournament as proxied by the probability implicit in the betting odds. This suggests estimation of the following equation:

$$\text{WINEQUALS1} = b_0 + b_1\text{IMPLIEDPROB} + e\,, \tag{11.4}$$

where WINEQUALS1 is a dichotomous variable equal to one if the golfer wins the tournament in question, and IMPLIEDPROB is the probability of winning that is implied in the posted odds. If the odds predict the related probabilities perfectly, then b_1 should equal one. However, due to the dichotomous nature of the dependent variable the error term will not be zero. This suggests use of a Logit estimation model, and the first column of Table 11.6 lists the results.

The regression coefficients from a Logit procedure are difficult to interpret directly. The sign of the coefficient of IMPLIEDPROB is correct, and the estimate is statistically significant. The equation itself is significant also, as evidenced by the Chi-squared value of 64.29 which would randomly be this high with a probability of less than 0.0001. The results of Eq. (11.4) are consistent with the proposition that the probabilities imbedded in the posted odds can help to predict the winner. However, it is hard to know what, if anything, to make of a regression when the dependent variable has 36 ones and 1,032 zeroes.

The main test of whether the implied probabilities are unbiased predictors is to see if adding the skills and performance variables can add to the predictive ability. Therefore, consider the following equation which adds the other potential independent variables as explained above in Eq. (11.2):

Table 11.6 Regression coefficients (t-statistics in parentheses).

Equation	(11.4)	(11.5)	(11.6)
Sample	All n = 1068	All n = 971	All n = 971
Method	Logit.	Logit	O.L.S.
Dependent Variable	WINEQUALS1	WINEQUALS1	IMPLIEDPROB
constant	-4.21*** (-17.42)	32.93 (0.95)	-0.04054*** (-4.62)
IMPLIEDPROB	8.633*** (8.33)	16.51** (2.08)	
DRIVDIST		-0.0613 (-0.99)	0.000419*** (3.655)
DRIVACC		-6.507 (-0.89)	-0.00857 (-0.60)
GIR		6.193 (0.59)	0.00894 (0.62)
PUTTPER		-12.90 (-0.61)	0.1157*** (2.67)
BIRDIEAVERAGE		0.473 (0.36)	-0.00523*** (-2.67)
PARBREAKERS		2.995 (0.07)	0.538*** (6.10)
RECENTFIRSTS		-0.184 (-0.92)	0.0128*** (23.32)
CAREERFIRSTS		-0.0233 (-0.35)	0.00197*** (11.36)
RECENTTOP10		0.0362 (0.65)	0.000152 (0.89)
Adjusted R^2			.744

Note: *, **, and *** indicate significance at the 10%, 5%, and 1% levels.

$$\begin{aligned}\text{WINEQUALS1} = {} & b_0 + b_1 \text{ IMPLIEDPROB} + b_2 \text{ DRIVDIST} + \\ & b_3 \text{ DRIVACC} + b_4 \text{ GIR} + b_5 \text{ PUTTPER} + \\ & b_6 \text{ BIRDIEAVERAGE} + b_7 \text{ PARBREAKERS} + \\ & b_8 \text{ RECENTSFIRSTS} + b_9 \text{ CAREERFIRSTS} + \\ & b_{10}\text{RECENTTOP10} + e . \qquad (11.5)\end{aligned}$$

The results of this equation are in the second column of Table 11.6. As the table shows, the efficient markets hypothesis stands up well. The Chi-squared value is 20.61, with the associated probability of only 0.024. The only significant variable is the probability implied in the odds. All of the skill variables and past performance variables essentially add nothing to the explanation. In regressions not shown, based on splitting the sample by the presence or absence of Tiger Woods, the same pattern emerges. The probability implied in the odds is the only statistically significant explanatory variable, and the equation predicts only a small part of the variation.

Finally, to illustrate that the set of skills and past performance variables is capable of measuring something meaningful, consider Eq. (11.6).

$$\begin{aligned}\text{IMPLIEDPROB} = {} & b_0 + b_1 \text{ DRIVDIST} + b_2 \text{ DRIVACC} + \\ & b_3 \text{ GIR} + b_4 \text{ PUTTPER} + b_5 \text{ BIRDIEAVERAGE} + \\ & b_6 \text{ PARBREAKERS} + b_7 \text{ RECENTSFIRSTS} + \\ & b_8\text{CAREERFIRSTS} + b_9\text{RECENTTOP10} + e . \\ & \qquad (11.6)\end{aligned}$$

This equation estimates how well the set of skills and past performance variables does predict the actual probability implied in the odds. The result, in the third column of Table 11.6, shows that this equation works very well. Roughly three quarters of the variation in the implied probabilities is explained by these variables. This result underscores the previous results. The skills can predict the odds, the odds can predict the winner, albeit to a small degree, and the odds seem to be an unbiased predictor of the winner in the sense that the skills, when added to the equation, are statistically insignificant and add no extra predictive power.

11.5 Summary

Two types of market efficiency are examined in this chapter. One is concerned with the ability of betting odds to capture in an unbiased way the best prediction of the outcome of a golf tournament. If a bias in the odds is uncovered, a betting algorithm leading to positive profit might be possible. This chapter finds no bias in regression equations for either the rank order of finish or for predicting the actual winner of the tournament. After controlling for the information in the odds, the extra variables do not increase the predictive power, therefore, they supply no wedge upon which to exploit the posted odds for a profitable betting system.

There is, perhaps, a finding of a tendency for the casino to offer odds that are too high on the field. A portfolio of bets on the field would have earned a handsome return in the 34 tournaments that offered a field bet in 2002. There is not much in the way of statistical analysis that can be done on the field bet. By definition, there are no individual statistics or past performances that can be used as regressors, because the field typically consists of about 114 different golfers each week, other than the 30 or so that the casino offers specific odds on. One illuminating calculation can be made. Given the odds on the field bet, and assuming that each individual golfer in the field has an equal probability of winning, the implied odds for each individual in the field can be calculated. For a 114-member field at even odds as in Table 11.1, the payoff is essentially 227/1 for each golfer. Since the highest odds offered on an individual golfer is usually 60/1 or 50/1, it would appear that the payoff for the field bet is at a much higher odds than for the listed longshots, however, you have to bet on all 114 of them.

Interestingly, other researchers have uncovered a similar favorite-longshot bias. However, it is not clear to this author whether the analogy holds. In one sense, betting on the field is like betting on a favorite at low odds. It is in this sense that the golf bet has a favorite-longshot bias similar to horse racing where favorites are underbet (leading to higher than justified odds in the parimutuel system) and longshots are overbet (forcing the odds down in the parimutuel system). In another sense, however, by betting the field one is actually betting on 114 separate longshots, albeit you have to bet on all of them, and only one can win. In this sense, the longshots actually win more often than the implied odds, which is the opposite of the usual favorite-longshot bias. Whether it is better to view a bet on the field as a bet on a favorite or as a tie-in sale of 114 longshot bets will be left for future golf-betting economists to ferret out.

A second type of market efficiency has to do with the decrease in transaction cost/profit margin as the volume of transactions increases. Comparing regression results from samples including and excluding Tiger Woods does not reveal any tendency for this type of efficiency. Furthermore, the implied percentage take of the casino is not lower in the tournaments in which Tiger is competing. If anything, the casino's profit margin is higher in such tournaments. An alternative hypothesis is that the betting public's demand is more inelastic when Tiger is competing. Thus, following the prediction of the textbook example of a uniform price setting monopolist, the implied price of a bet would be higher.

Unfortunately, readers of this chapter will not be getting rich anytime soon by handicapping professional golf. The gambling odds market seems to be efficient in the sense of capturing all of the relevant available information, at least with respect to skills and past performance. Still, only a very small portion of the variation in the outcomes is explained by the odds. This may be a testament to just how difficult the game of golf is. However, the possibility is left open that other variables, not included here, might have some predictive power over and above that given in the odds. If so, perhaps a betting algorithm could be developed. Variables that could be tried might focus more on recent finishes (to measure a hot hand tendency), expected weather patterns including windiness (some players may be systematically better in the wind, rain, or heat) or past success on a particular type of golf course (some favor long hitters, others favor accuracy and putting, etc.). The data requirements for such a study seem formidable but not insurmountable, and I encourage future researchers to take up this challenge.

Chapter 12

Still Looking for Economic Impact: The Case of the Golf Majors[1]

Economic impact studies are carried out in a variety of settings by a variety of people. The settings include things as varied as building a mass transit system, relocating an airport, implementing a new regulation, building a sports stadium, hosting an Olympic Games, or hosting a Major golf tournament. The people who desire the studies and carry them out include corporate executives contemplating business decisions, politicians, other public policy makers, taxpayer associations, consultants, and voters. Studies are also carried out for a variety of purposes. On the one hand it is natural for policy makers to try to foresee the results of any particular decision that they have to make, because they want to make the best decision. Presumably, the closer a forecast can come to predicting the actual results, the better the decision maker can decide among the available options. On the other hand, proponents or opponents of a particular project may have already decided on the decision they want. They will desire an impact study to confirm their point of view and convince others of the desirability, or lack thereof, of the project in question.

There are also a variety of ways to go about performing an economic impact analysis. These can be divided into two basic methodologies which

[1]This chapter draws on material originally published in Shmanske, S. (2012a) *Handbook on the Economics of Mega Sporting Events* eds. Maennig, W. and Zimbalist, A., Chapter 25 "The Economic Impact of the Golf Majors," (Edward Elgar Publishing Ltd, Northampton, Massachusetts) pp. 449-460.

can be further subdivided. The prospective, or *ex ante*, method is forward looking and attempts to forecast or predict the future outcome. Alternatively, *ex post* methods look at a decision after the fact to determine what actually did happen. Digging deeper, the data for these analyses can come from guesses, both wild and educated, extrapolations from surveys about either expected or actual expenditures, and published data on sales taxes, employment, payroll, and the like. Obviously, *ex post* studies cannot help beforehand to make any particular decision, but they do help in assessing the accuracy of any previous *ex ante* study's predictions, and in that way, can influence future decisions by providing a critical analysis of an *ex ante* study's methodology.

This chapter will summarize the methodology and results of two *ex ante* studies of the potential economic impact of hosting a Major golf tournament, and carry out a multiple year and multiple site *ex post* econometric study of the employment effects of hosting Major golf tournaments. The next section provides some background about the Major golf tournaments and looks in particular at studies of the 2008 and 2009 U.S. Open Championships and the underlying methodology that they use. Following that, I will introduce the data and methodology used in my research. The last section presents the results and summarizes what we can and cannot learn from looking at the statistics.

12.1 Background

Professional golfers compete in tournaments year round and all over the world. Traditionally, four tournaments have attained a special status, becoming known as the Majors. They are: The Masters, played each year at the same course in Augusta, Georgia; The U. S. Open Championship, which rotates among different courses in the United States each year; The Open Championship or "British Open," which rotates among different courses in the British Isles each year; and the Professional Golfers Association (PGA) Championship, which rotates among different courses in the United States each year. Although these tournaments do not near the magnitude of the NFL's Super Bowl, among the community of golfers and golf fans they are anticipated with equal fervor. The golf Majors have among the largest purses, and carry extra monetary significance for the players because of endorsement possibilities for the winners and because of the special future playing rights afforded to the winner and top finishers. Because of the playing exemptions, the money, and the prestige, the Majors attract the top talent, consequently attracting large crowds of spectators.

The sporting significance of the golf Majors is undeniable, but quantifying the economic significance for the local community is not nearly as straightforward. Sports boosters will always point to the additional spending by visitors who come to attend sports competitions. There is also extra spending by those staging an event in terms of lighting and power, crowd control, parking, and readying the course beforehand. Some of this spending takes place weeks or even months in advance of the event itself. This extra demand for local goods and services increases income in the local community and may lead to further increases in future spending, thus having a "multiplied" effect in the local economy. Thus, the typical *ex ante* study, for example, estimates the expenditure flows associated with hosting the tournament and multiplies this expenditure to account for additional spending undertaken by the recipients of the original expenditures.

For example, a local association forecast the impact of the 2009 U.S. Open held at Bethpage Black on Long Island.[2] This report estimates the direct spending by the U.S. Golf Association (USGA), the sponsor of the tournament, and its vendors to be $24.5 million. Using multipliers from the RIMS II[3] input-output model of the Long Island economy, the direct spending leads to indirect spending of $23.5 million for a total of $48 million even before spending by the participants and spectators. The USGA itself predicts that its gross revenues from tickets, concessions, logo'ed merchandise, and the like to be $37.4 million. Albeit most of this revenue stream does not stay in the local community, but close to nine percent skimmed in sales taxes does remain. And these expenditures are in addition to local lodging, food and transportation expenditures by attendees.

[2]See Kamer, P. M. (2009). The 2009 U.S. Golf Open at Bethpage Black: Its Impact on the Long Island Economy, Research Report from the Long Island Association.

[3]The RIMS II models are developed by the Bureau of Economic Analysis of the U.S. Commerce Department and are accessible at: http://www.bea.gov/bea/regional/rims/.

To get a feel for the expenditures in the local economy by attendees consider a report done on the economic impact of the 2008 U.S. Open Championship at Torrey Pines Golf Course in San Diego.[4] Using surveys of spectators to get demographic characteristics and spending patterns this report estimates direct expenditure of spectators at $73.6 million. Again, multipliers from the RIMS II model, this time for Southern California, were used to estimate additional indirect expenditures of $68.5, for a total of over $142 million.

Supposing that the experiences in 2008 and 2009 were similar we can get a ballpark estimate of the totals that are expected. From Bethpage Black we get the direct expenditures of the tournament promoters ($24.5 million), and the indirect expenditures that they cause ($23.5 million). From Torrey Pines we get direct expenditures of spectators which probably include the $37.4 million paid to the USGA and will not be double counted here ($73.6 million), and the indirect expenditures they cause ($68.5 million). Throw in some expenditures by the participants and some sales taxes that get repatriated to the local community for a total of nearly $200 million. To put this in perspective, similar *ex ante* methodology has been used to predict the economic impact of hosting a Super Bowl at about $500 million. But also keep in mind that by looking at actual data after the fact *ex post* studies of the Super Bowl reveal an actual effect of about $50 million or roughly one-tenth as much.[5]

The analyses of the 2008 and 2009 U. S. Opens notwithstanding, there are many critics of the basic methodology of adding up "direct" expenditure and applying an input-output table to estimate additional "indirect" expenditures. Each step in the methodology can be criticized. For example, consider one of the direct expenditures, spending in restaurants by spectators. One cannot simply look at the total restaurant receipts for the week, because this would be assuming that the restaurants would otherwise be empty. Furthermore, even looking only at the spending by spectators, perhaps through a survey, would lead to an overestimate for

[4]2008 U.S. Open Economic Impact Analysis, San Diego State University, Center for Hospitality and Tourism Research.

[5]See, Matheson, V. A. (2012) *The Oxford Handbook of Sports Economics, Volume 1: The Economics of Sports*, eds. Kahane, L. H. and Shmanske, S., Chapter 24 "Economics of the Super Bowl," (The Oxford University Press, London) pp. 470-84.

at least two reasons. First, a visitor, even one who attends the tournament, may have visited the locale anyway even if the tournament was not scheduled. This visitor's spending should not be attributed to hosting the tournament, because it would have happened anyway. Along the same lines, perhaps the visit was timed to catch the golf tournament but was shifted from another time. The visitor's extra spending on the week of the tournament looks good, but is offset by the lack of his spending on the other week when he would have come instead. Second, the tourist's visit to a restaurant may have crowded out a local resident's visit. Consider a restaurant that is always filled on weekends. During the tournament it is filled with tourists as the regulars stay home to avoid the added crowds. By surveying the expenditure by tourists on restaurant meals but not offsetting the lost expenditure by local residents, an overestimate of the additional direct spending is obtained. These problems are ameliorated somewhat by looking at a time series of expenditure on weeks including and not including the special event. Only the upward blip in spending, if there is one, should count as being caused by hosting the tournament. But when one surveys visitors to ask the amount they spend, this adjustment is apparently not being made.

To summarize so far, some of the direct expenditures are not additional, and would have taken place anyway, in the local economy, but perhaps elsewhere in place or time. And, some "new" direct expenditures may displace or crowd out other local expenditures. It is not a stretch to say that the direct expenditures due to hosting an event like a Major golf championship are often overestimated.

But there is a further problem. The multiplication of these direct expenditures to count up the additional rounds of extra spending can also be criticized. For the Bethpage Black study the direct expenditures of $24.5 million led to additional expenditures of $23.5 million so the multiplier was 0.959. For the 2009 Open at Torrey Pines the direct spending of $73.6 million led to additional spending of $68.5 million for a multiplier of 0.931. To uncover the potential problem, however, requires more explanation of how these multiplier models work in theory.

Consider a simple model in which a consumer spends all his income divided into only two categories: hospitality (including meals and lodging) and retail sales. Suppose the average consumer spends 40% on hospitality, and 60% on retail sales.[6] How much of this spending normally gets recycled in the local economy? Suppose that typically, for each dollar of spending on hospitality, \$0.60 goes to local citizens and \$0.40 "leaks" out of the local economy as income and profits to non local owners such as corporate executives in corporate headquarters and nationally and even internationally dispersed stockholders. Finally, suppose the corresponding breakdown for retail sales is 50-50. Now consider what happens when an out of town visitor spends \$40 on hospitality and \$60 on merchandise for a total of \$100 of new, direct spending. See Table 12.1.

Table 12.1 Illustrative multiplier analysis.

Sector	New direct spending	Leakage from direct spending	First round of secondary spending	Leakage from first round	Second round of secondary spending	Total all rounds of secondary spending
Hospitality	40	16	21.6	8.64	11.66	46.96
Retail	60	30	32.4	16.2	17.50	70.43
Total	100	46	54	24.84	29.16	117.39
		\	/	\	/	
		= 100		= 54		

In Table 12.1 the first column of figures is the new direct spending in the local economy distributed between the two sectors. The second column indicates by sector how much of that spending does not stay in the local economy, namely 40% in the hospitality sector and 50% in the retail sector. This leakage of 46 means that the total income that stays in the local economy is 54 which is spent between the two sectors in the usual 40%-

[6]Note that all the baseline numbers and percentages used in this illustrative analysis are made up for ease of calculation. The actual numbers would be based on longer term analysis of the local economy.

60% mix. Thus, the third column of figures indicates secondary spending of 21.6 in the hospitality sector (that is, 40% of $54) and 32.4 in the retail sector. As in the first leakage column, 40% of hospitality spending leaves the community (going to corporate headquarters, for example) and 50% of the retail spending leaks out (going perhaps to out of town wholesalers in the cost of goods sold category). The total leakage from the first round is 24.84, leaving 29.16 to be spent again according to the usual percentages.

This process goes on indefinitely, however, in each new round of additional spending the amounts are reduced by 46%. The total summation of all the infinite rounds of spending and respending amounts to $117.39. Since the new direct spending in the example was $100, the multiplier in this example is 1.1739. This is how the multiplier is estimated for the local economy, but it is most probably an overestimate of the actual effect as the following will argue.

The problem with applying the usual multiplier to the new extra spending during the golf tournament is that the new spending most probably does not fit the usual pattern for the community. In particular, in the retail sector, much of the spending goes directly to the USGA for their merchandise so a greater proportion leaks out of the local economy in the first round. In the hospitality sector, it would be usual for the local hotels to exploit the increased demand by raising prices (or eliminating discounts) and funneling the extra funds received back to corporate headquarters and not to the local labor force in increased wages. So for the first round of spending only, consider what happens if the leakages increase from 40% and 50% to 50% and 60% for the hospitality and retail sectors respectively. The results are in Table 12.2.

We now see that the spending totals are lower. This is due solely to the extra leakage from the initial spending as shown in the second column of figures. Since more of the original spending leaves the community there is less to filter through the local economy in the rounds of secondary spending. The total of all rounds of secondary spending now adds up to $95.65 for an implied multiplier of approximately 0.96 which is nearly 20% lower than the unadjusted case in Table 12.1.

The bottom line is that the new direct spending can easily be overestimated and that the multiplier applied is probably too high. Both these errors lead to an overestimate of the economic impact of staging a

Table 12.2 Illustrative multiplier analysis (revised).

Sector	New direct spending	Leakage from direct spending	First round of secondary spending	Leakage from first round	Second round of secondary spending	Total all rounds of secondary spending
Hospitality	40	20	17.6	7.04	9.50	38.26
Retail	60	36	26.4	13.2	14.26	57.39
Total	100	56	44	20.24	23.76	95.65
		\ = 100	/	\ = 44	/	

special event like a Major golf championship. Sometimes proponents of the use of this methodology attempt to make adjustments for these effects but more often than not, boosters will uncritically accept the rosiest of unadjusted forecasts in advocating for local political and economic support for a prospective event.

One could probably get a near unanimous consensus that the rosiest estimates coming from *ex ante* analyses are overstated. This is because the rhetoric surrounding the rosiest estimates comes from project supporters who are not disinterested in the outcome of a decision to host a Major golf tournament. However, this does not mean that there is no net local economic spillover benefit. Thus, it becomes an empirical question to measure the existence and extent of the local economic impact of hosting such an event. Unfortunately, it is not an easy question even from an *ex post* perspective informed by the data. It is certainly much more involved than an uncontrolled year-to-year comparison of economic activity as Victor Matheson humorously points out:

> . . . the NFL reported that, "Thanks to Super Bowl XXXIII, there was a $670 million increase in taxable sales in South Florida compared to the equivalent January-February period in 1998."

> (NFL Report, 1999) . . . Unfortunately for the NFL, their study is woefully inept as the league neglected to account for factors besides the Super Bowl, such as inflation, population growth, and routine economic expansion, that could account for the rise in taxable sales. As noted by Baade and Matheson (2000), over 90% of the increase can be accounted for by these variables. . . . Finally, it is worth noting that taxable sales in the area during January-February 2000, the year after the game, were $1.26 billion higher than in the same months during the Super Bowl year. Strangely, the NFL never publicized a story announcing, "Thanks to the lack of a Super Bowl, there was a $1.26 billion increase in taxable sales in South Florida compared to the equivalent January-February period in 1999." [Matheson, 2012]

The unsophisticated comparison that the NFL trumpets cries out for a more systematic, more inclusive look at the data. Ideally, comparisons of the level of economic activity should be made for multiple years and for multiple locations. A study of the American rotating golf Majors, namely the U.S. Open Championship and the PGA Championship, allows undertaking exactly this type of analysis. In a multiple regression setting with fixed yearly effects and fixed county effects, we can isolate the effect on the local economy in the year of (and the years following) the staging of a Major golf tournament. This is precisely the goal of this chapter.

12.2 Models and Data

Consider Fig. 12.1 which illustrates several possible patterns of economic activity, measured here as payroll, associated with hosting a mega event. In each panel year T represents the treatment year in which the golf Major was staged. Up until T there is a baseline of economic activity. For ease of illustration this baseline is pictured as stationary but in the fixed effects model to be estimated below, the baseline will follow whatever average trend in the data actually exists. In Panel A there is a one-year increase associated with the event. This effect would be captured in a regression model by the inclusion of a dummy variable which would single out the treatment year in question. But if hosting a Major golf tournament had a lasting effect as illustrated in Panels B or C, the one-time dummy variable would fail to pick it up.

Panel B of Fig. 12.1 illustrates the possibility that there is a lasting effect of hosting a Major. For example, the publicity and good will associated with hosting a Major could lead to a permanent shift in the level of economic activity in the area. This effect could be picked up by the inclusion of an indicator variable that separated the post treatment economic activity from that occurring before the event. The indicator variable would take on the value of zero in the years before the event and the value of one for the year of the event and the years following. Alternatively, a difference-in-differences model can also capture the effect with a single dummy variable for the treatment year. If the dependent variable is the year to year change in economic activity, then the dummy variable for the treatment year stands out differently from the rest of the time series because it captures the change in the year to year change. However, formulating the dependent variable as the year to year change does come with the cost of reducing the number of observations.

Panel C of Fig. 12.1 illustrates the case in which a one time shift is not the proper specification because the event causes a lasting increased trend

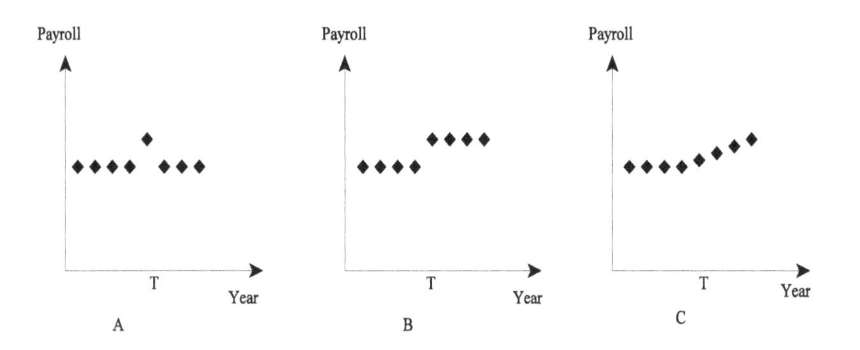

Fig. 12.1 Patterns of economic activity

in economic activity above the baseline. In such a case a post-treatment trend variable must be added to properly capture the effect.[7]

Ultimately, the patterns in Panels A-C can be combined. A change in activity due to hosting a Major golf tournament could take the form of a one time or lasting shift with or without a lasting increased trend. Indeed, any pattern could follow the hosting of a mega event. With limited data, one must impose a specific functional form in a regression model to test whether such a pattern explains the data better than the null hypothesis that hosting the event has no effect. The beauty of looking at multiple Majors in different locales over multiple years is that it allows testing for the existence of any pattern of economic activity by the inclusion of indicator variables tied to the number of years following the hosting of the event itself. One need not impose a conjectured pattern beforehand. A description of the data will make this point clearer.

Counting both the U.S. Open and the PGA Championship, for the years 1991 to 2008 there were 36 tournaments contested on a total of 24 different golf courses. These 24 locales and 18 years, listed in Table 12.3, form the data set for the analysis. Payroll data by year for each of the relevant counties were retrieved from the County Business Patterns data available online. Yearly data are used to get the closest comparison across the different tournaments and to avoid picking up a spiking of economic activity around the event that is ultimately offset by a lull in activity elsewhere in the year. If there is a large enough true gain to the local economy it should persist and show up in the yearly data.[8]

Two dependent variables will be explored. The first, total annual payroll, is a good measure of overall economic activity in a county. This measure should pick up any growth in employment or wages due to added economic activity. It will also pick up gains in part time, overtime, or

[7]See Hagn and Maennig [2008], and Hotchkiss, *et al.* [2003] for examples of attempts to capture the type of effect illustrated in Panels B and C.

[8]. . . Or maybe not. If the $200 million figure mentioned in the introduction is the correct magnitude, its influence may be swamped. It is a lot of money, but in many of the counties involved it represents less than one percent of the total annual payroll. Averaging out the fixed effects for each county will put all counties on the same scale, but still, a $200 million amount will be large qualitatively in some counties while seeming like random measurement error in others.

Table 12.3 U. S. Open and PGA Championships 1991-2008.

Year	Sponsor	Golf Course	City, State	County
1991	USGA	Hazeltine National Golf Club	Chaska, MN	Carver
1992	USGA	Pebble Beach Golf Links	Pebble Beach, CA	Monterey
1993	USGA	Baltusrol Golf Club	Springfield, NJ	Union
1994	USGA	Oakmont Country Club	Oakmont, PA	Allegheny
1995	USGA	Shinnecock Hills Golf Club	Southampton, NY	Suffolk
1996	USGA	Oakland Hills Country Club	Bloomfield Hills, MI	Oakland
1997	USGA	Congressional Country Club	Bethesda, MD	Montgomery
1998	USGA	The Olympic Club	San Francisco, CA	San Francisco
1999	USGA	Pinehurst Resort & Country Club	Village of Pinehurst, NC	Moore
2000	USGA	Pebble Beach Golf Links	Pebble Beach, CA	Monterey
2001	USGA	Southern Hills Country Club	Tulsa, OK	Tulsa
2002	USGA	Bethpage State Park	Farmingdale, NY	Nassau
2003	USGA	Olympia Fields Country Club	Olympia Fields, IL	Cook
2004	USGA	Shinnecock Hills Golf Club	Southampton, NY	Suffolk
2005	USGA	Pinehurst Resort & Country Club	Village of Pinehurst, NC	Moore
2006	USGA	Winged Foot Golf Club	Mamaroneck, NY	Westchester
2007	USGA	Oakmont Country Club	Oakmont, PA	Allegheny
2008	USGA	Torrey Pines Golf Course	San Diego, CA	San Diego
1991	PGA	Crooked Stick Golf Club	Carmel, IN	Hamilton
1992	PGA	Bellerive Country Club	St. Louis, MO	Saint Louis City
1993	PGA	Inverness Golf Club	Toledo, OH	Lucas
1994	PGA	Southern Hills Country Club	Tulsa, OK	Tulsa
1995	PGA	Riviera Country Club	Los Angeles, CA	Los Angeles
1996	PGA	Valhalla Country Club	Louisville, KY	Jefferson
1997	PGA	Winged Foot Golf Club	Mamaroneck, NY	Westchester
1998	PGA	Salhalee Country Club	Redmond, WA	King
1999	PGA	Medinah Country Club	Medinah, IL	Du Page
2000	PGA	Valhalla Country Club	Louisville, KY	Jefferson
2001	PGA	Atlanta Athletic Club	Duluth, GA	Gwinnett
2002	PGA	Hazeltine National Golf Club	Chaska, MN	Carver
2003	PGA	Oak Hill Country Club	Rochester, NY	Monroe
2004	PGA	Whistling Straits	Kohler, WI	Sheboygan
2005	PGA	Baltusrol Golf Club	Springfield, NJ	Union
2006	PGA	Medinah Country Club	Medinah, IL	Du Page
2007	PGA	Southern Hills Country Club	Tulsa, OK	Tulsa
2008	PGA	Oakland Hills Country Club	Bloomfield Hills, MI	Oakland

temporary added employment, that may be necessary to deal with special problems posed by staging a mega event. Total annual payroll should also pick up gains coming from the indirect spending part of the story. For example, The U.S. Open is held in June, potentially creating a bump up in economic activity that will start to filter through the rest of the economy for the rest of the year.

Second, this paper looks at the payroll in the hotel/motel sector of the economy which should be especially sensitive to the influx of visitors coming for the mega event.[9] These visitors include contestants and spectators, who may need housing for the week of the event, and also employees of the sponsoring association and its out of town vendors who may visit weeks or months in advance of the event. Keep in mind that the direct spending for temporary housing is only part of the direct expenditure stream associated with the event, so the coefficient estimates should be appropriately smaller when the second dependent variable is used.

The following equation is estimated with O.L.S. using the robust standard errors sub-procedure.

$$\text{PAYROLL} = a_0 + A_1\text{MAJOR} + A_2\text{YEAR} + A_3\text{COUNTY} + e\,. \quad (12.1)$$

Each observation of PAYROLL refers to a given county's payroll in a given year measured in constant 2008 \$US.[10] As discussed above, PAYROLL is either the total yearly payroll or the yearly payroll in the hotel/motel sector. The A's are vectors of coefficients associated with the matrices of data.

The main variables of interest are in the matrix, MAJOR, which is composed of vectors of dummy variables denoted, M_i, for i=0-4. M_0 is the

[9]The County Business Patterns data are broken down by NAICS (formerly SIC) industry codes allowing examination of particular sectors of the economy.

[10]The regressions are done on payroll levels. Logarithms of earnings are often used in salary regressions to correct for the heteroskedasticity usually present in individual salary regressions, but are not appropriate in the present context. Using levels, the regression seeks to find a fixed dollar addition to economic activity in the county-year in question because the tournaments themselves are roughly the same size regardless of the size of the county in which they are staged. Using logarithms would be testing for a fixed percentage effect across counties for which expectation there would be no basis in theory. Thus, a regression using logarithms of payrolls as the dependent variable would be theoretically biased against rejecting the null hypothesis of no effect. As a robustness check, such a regression was run and, as expected, returned insignificant coefficients.

dummy variable equal to one for the treatment year, that is, for an observation where the county hosted a Major in the year in question. Its coefficient will allow us to test the hypothesis that hosting a Major golf tournament is good for the local economy in the year in question, at least in the sense of its effect on payrolls, against the null hypothesis that hosting a Major golf tournament has no contemporaneous effect. The variables M_1 through M_4 pick up the effects in the first through fourth years following the hosting of the tournament. Using the five variables, M_0-M_4 allows for any pattern of effects on economic activity coincident with and following the hosting of a Major golf tournament.[11]

The matrix, YEAR, is composed of dummy variable vectors which isolate the effect of each year. The pattern of these fixed effects coefficients in A_2 should coincide with the general level of economic activity for the year.

The matrix, COUNTY, is composed of dummy variable vectors which isolate the effect of each locale. The pattern of these county effects coefficients in A_3 should coincide with the overall size (or industry specific size) of the county. Some interesting comparisons should show up in the coefficients. For example, consider Monterey County in California, which is small compared to many of the counties in the data, but which is a resort area with a large hospitality industry sector. When overall payroll is used, Monterey County will be below average in magnitude, but when payroll in the hotel/motel sector is used, Monterey will be above average in size.

To avoid singularity in the estimation, one county and year must be left out of the fixed effects, to be picked up in the constant term, a_0. Therefore, the dummy variables for Allegheny County, which includes the Oakmont Country Club in Oakmont, Pennsylvania, and for 1991, the first year of data, are omitted. Consequently, the yearly fixed effects represent the differences between the year in question and 1991. Meanwhile, the county effects measure differences between the county in question and Allegheny County.

Summary statistics for the two dependent variable payroll measures are listed in Table 12.4. Useable figures for the hotel/motel sector are not

[11]Experimentation with other time periods, including looking only at M_0, the contemporaneous effect, did not alter the inferences.

available for all counties for all years so the data set is correspondingly smaller. There are no omissions for the overall county payroll.

Table 12.4 Summary statistics.

Variable	Mean	Standard Deviation	Minimum	Maximum	n
Total county payroll (2008 $US)	28.4 Billion	37.6 Billion	574 Million	187.4 Billion	432
Hotel/motel county payroll (2008 $US)	210 Million	278 Million	641,000	1.1 Billion	370

12.3 Results

The statistical results are listed in Table 12.5. Unfortunately for the boosters of the golf Majors, I was unable to identify any significant increase in county payrolls timed with hosting a U. S. Open or a PGA Championship. The coefficients of the main variables of interest, M_0-M_4, are statistically insignificant in both regressions. The problem is not due to multicollinearity. In regressions that include only M_0, its coefficient is statistically insignificant in both regressions and the point estimate is even negative for the case of total county payroll.[12]

There does not appear to be any problem in the data or the calculations. The overall fit is good in both equations. The use of robust standard errors did not change any of the inferences that would have been drawn from O.L.S. Given the good fit and the lack of evidence of any severe problem with heteroskedasticity, additional specification searches were foregone.

With respect to the general trends in the data, the pattern of the fixed effects coefficients tracks what we know about the U. S. macro economy. There were slight, relatively short recessions in the early 1990's and in the

[12]The F-tests for inclusion of all five M_i variables are insignificant at the 0.10 level in both equations.

Table 12.5 Coefficient estimates (t-statistics).

Dep. Var.	Total Payroll (1000 $US2008)	Hotel/Motel Payroll (1000 $US2008)
Constant	2.40E7 (21.05)***	104,392 (7.81)***
M_0	40,162 (0.06)	4968 (0.62)
M_1	43,545 (0.06)	5764 (0.65)
M_2	381,747 (0.54)	9898 (1.15)
M_3	443,625 (0.74)	2197 (0.29)
M_4	246,317 (0.46)	1254 (0.12)
1992	-461,138 (-0.31)	-8796 (-0.61)
1993	-641,034 (-0.40)	-8918 (-0.61)
1994	-489,552 (-0.30)	-9836 (-0.64)
1995	303,292 (0.20)	-8064 (-0.56)
1996	1,211,624 (0.87)	-2046 (-0.16)
1997	2,364,686 (1.85)*	4212 (0.35)
1998	4,347,068 (3.65)***	17,820 (1.56)
1999	5,885,534 (4.75)***	31,931 (2.60)***
2000	7,487,080 (5.26)***	55,157 (3.33)***
2001	7,008,526 (5.32)***	31,080 (2.80)***
2002	5,911,673 (4.85)***	18,562 (1.77)*
2003	6,021,792 (4.82)***	19,903 (1.73)*
2004	6,315,584 (4.96)***	29,790 (2.66)***
2005	6,446,777 (4.94)***	35,181 (3.00)***
2006	7,445,771 (4.94)***	46,619 (3.08)***
2007	7,832,549 (4.95)***	45,026 (2.87)***
2008	7,359,993 (4.84)***	43,098 (2.50)**
Carver	-2.71E7 (-33.87)***	-120,179 (-8.06)***
Cook	8.86E7 (50.96)***	639,657 (25.89)***
Du Page	-736,399 (-1.49)	-3858 (-0.33)
Gwinnett	-1.72E7 (-44.62)***	-91,886 (-6.21)***
Hamilton	-2.50E7 (-38.40)***	-127,973 (-9.91)***
Jefferson	-1.33E7 (-23.12)***	-56,350 (-4.93)***
King	2.31E7 (13.37)***	184,681 (14.88)***
Los Angeles	1.39E8 (50.36)***	828,920 (33.96)***
Lucas	-2.03E7 (-25.48)***	-103,162 (-9.26)***
Monroe	-1.32E7 (-14.76)***	-82,095 (-6.87)***
Monterey	-2.45E7 (-33.15)***	35,140 (3.09)***
Montgomery	-8.85E6 (-20.69)***	-41,395 (-3.92)***
Moore	-2.69E7 (-23.32)***	-91,754 (-5.90)***
Nassau	-3.90E6 (-6.90)***	-60,707 (-5.31)***

Table 12.5 Coefficient estimates (t-statistics) (continued).

Dep. Var.	Total Payroll (1000 $US2008)	Hotel/Motel Payroll (1000 $US2008)
Oakland	7.59E6 (9.22)***	-45,915 (-3.94)***
Saint Louis City	-1.60E7 (-22.62)***	-43,480 (-3.54)***
San Diego	1.42E7 (8.64)***	476,283 (20.59)***
San Francisco	3.77E6 (5.71)***	509,764 (26.36)***
Sheboygan	-2.63E7 (-31.15)***	-108,754 (-9.79)***
Suffolk	-6.27E6 (-15.80)***	-69,449 (-5.69)***
Tulsa	-1.65E7 (-27.70)***	-83,365 (-7.18)***
Union	-1.67E7 (-25.22)***	-93,268 (-7.63)***
Westchester	-7.84E6 (-18.19)***	-27.369 (-2.20)**
R^2	.989	.977
n	432	370

Note: *, **, *** denote statistical significance of estimate at .10, .05, and .01 levels.

early 2000's, and the start of a major recession in 2008. These are all picked up in the results. Compared to the starting year of 1991, the coefficients show a decline in payrolls in 1992 through 1994, followed by strong growth in payrolls from 1995 to 2000. In 2001 and 2002 payrolls dipped again, then grew steadily through 2006 or 2007 before dipping again slightly in 2008.

Meanwhile, the county effects coefficients are of plausible magnitude and related, as they should be, to the relative size of the county. Compared to the omitted Allegheny County, counties like Los Angeles in California and Cook in Illinois are orders of magnitude larger, while others like Sheboygan in Wisconsin or Moore in North Carolina are significantly smaller. The regressions were even able to pick up the significant negative coefficient for Monterey County in California in the total payroll regression due to its small size, while still picking up a significant positive coefficient for hotel/motel payroll regression because of the resort nature of the county. The polar opposite of California's Monterey County is Michigan's Oakland County, the home of the Oakland Hills Country Club. Oakland

County is above average in size and has a significant positive coefficient in the total payroll regression, but has a small presence in the hospitality sector and, therefore, a significant negative coefficient in the hotel/motel payroll regression. Monterey and Oakland are the only two counties in the data with such sign reversals.

With respect to the main variables of interest, how can the lack of significant effects on payroll be reconciled with the obvious intuitive notion that people are attracted to an area to attend a Major golf championship event, and while there, spend money in the local economy? There are several possible ways to answer this question but perhaps crowding out and time shifting of demand are the most straightforward.[13]

For example, suppose the counties in the sample host a Major golf tournament once every dozen years or so. What happens in the other 11 years? One could imagine an area that hosts some sort of special festival (perhaps even a different one) every year. For one year it just happens to be the PGA Championship. When looked at in the data, the year with the golf Major does not stand out from the rest of the data. What has really happened is that the golf tournament has crowded out some other event that would have occurred in the year in question. So it is not all of the direct spending associated with the golf tournament that is added; it is only the amount above and beyond what would have happened that year anyway. Unfortunately it is impossible for any research design to pick up what would have happened in the absence of hosting a Major golf tournament. In any one year, an area either does, or does not host a Major and it is impossible to say what would have happened in the counter-factual case. Controlling through time and across different areas is the best that can be done. Hosting a tournament may be nice, but no one can ever know what would have happened that week if the tournament did not take place. The

[13]Sports economists have looked intensively at the crowding out issues at least as early as 1997, see Noll, R G. and Zimbalist, A. eds. (1997) *Sports, Jobs, and Taxes.* (Brookings Institution Press, Washington, DC). A recent excellent survey of empirical work is Coates, D. and Humphries, B. R. (2008). Do Economists Reach a Conclusion on Subsidies for Sports Franchises, Stadiums, and Mega-Events?, *Econ Journal Watch*, 5(3), pp. 294-315. The theoretical issues are explained thoroughly in von Allmen, P. *The Oxford Handbook of Sports Economics, Volume 2: Economics Through Sports*, eds. Shmanske, S. and Kahane, L. H. Chapter 18 "Multiplier Effects and Local Economic Impact," (The Oxford University Press, New York) pp. 321-334.

data seem to indicate that something of similar magnitude would have taken place in absence of the tournament and was crowded out that year and perhaps time shifted to another year in the data.

Notwithstanding the crowding out and the lack of the ability to document the effects of added direct expenditure and the indirect secondary expenditure, resident and nearby golf fans clearly benefit from the presence of the tournament. These people may seem incredulous that such an event does not cause a positive blip in economic statistics. But much of their gain might be captured as consumer surplus from the ability to attend the event and, thus, not show up in employment data or any other data. As a possible testament to this explanation consider that, in many cases, local resident golf fans volunteer to help stage the event. And in some cases there is a waiting list to volunteer or even a payment made to "volunteer" at the tournament. These people are clearly better off due to the presence of the golf tournament, but their gains do not show up in the data. Therefore, it follows that the desirability of hosting such an event need not be dependent upon showing a gain in local employment, income, or production. However, by the same token as the crowding out scenario in the preceding paragraph, these obvious gains for the golf fans should be offset by the consumer surplus lost by non golf fans who may lose the opportunity to do something else that week and who may suffer from added congestion. And in some cases it is even golfers themselves who lose out, as was the case when a local public-access, municipal golf course nearby the Olympic Club in San Francisco was closed to provide parking for the 1998 U. S. Open. Data on gains and losses in consumer surplus are simply not available, so the use of measurable series is the next best thing. When the measurable data on county payrolls are used, it appears that staging a Major golf tournament has no significant effect.

Chapter 13

The Best of the Rest

The previous chapters have highlighted my published and unpublished research on the economics of golf. As interest in the topic, and in sports economics in general, has grown, other researchers have started to use golf and golfers as the focus of their analyses or the source of their data. The imagination with which these other economists have pursued the topic has always impressed me. I have often noted to myself, "wish I had thought of that." In this chapter I briefly describe several of the lines of research that others have undertaken. This will involve setting the stage by describing the nature of the underlying theory that is tested, the nature of the data innovation that is employed, or the nature of the clever new empirical question that is explored. I cover in turn celebrity endorsement, incentives when competing against a superstar, the link between beauty and incentives, behavioral economics, and the new way that statistics are measured on the PGA TOUR.

13.1 Tiger Woods and Celebrity Endorsement

Many companies pay large sums of money, and brag about it, to have their products associated with sports stars and other celebrities. Immediately, two basic questions arise. From the point of view of the companies and their marketing departments the question is, "does it work in practice?" Economists are also interested in this question and have developed a clever method of shedding light on the situation. But for economists there is also a more basic question of, "why or how does it work in theory?" We will return to the practical question after introducing and exploring two views on the theoretical question of how celebrity endorsement is thought to work. The existing empirical work may not settle this second question, but I present the argument in the hope that future researchers will attack the issue with cleverness that escapes me.

So why does advertising work? The simple answer, and the starting point for Philip Nelson's seminal article,[1] is that advertising supplies information about the product or its terms of sale, including where, when, and how much. This type of information became known as "hard information," and it is not difficult to see why such information might be valuable to consumers and, therefore, increase sales.

But Nelson delved farther into the subject identifying another type of information, called "soft information," such as that embodied in images of cars going through puddles or around curves, or images of curvy, scantily clad, professional models having fun on the beach while holding cans of diet soda or light beer. There is really no information in such an advertisement other than the fact that the product or company exists and that the company chooses to spend money, and sometimes a great deal of money, on advertising. Celebrity endorsements of companies and products such as Michael Jordan sitting around in a tee shirt, Shaquille O'Neal sitting in a compact car, or Tiger Woods hitting golf balls on a practice range, seem to fall into this latter category of soft information. It seems to me that it is a legitimate question to ask why this type of advertising works.

An answer that I do not find particularly persuasive is based on a type of psychological persuasion. Supposedly, if I want to "be like Mike," I should eat food that Michael Jordan endorses, wear his brand of underwear, and wear his sneakers. Does anyone really think that this works? Seriously. Will all my friends look like professional Hollywood models if I pick the right brand of light beer? At a deeper, subconscious level, there may be something about an association made between the excellence of the

[1]See Nelson, P. (1974). Advertising as Information, *Journal of Political Economy*, 82(4), pp. 729-754.

on screen spokesperson or image and the excellence of the product or company. Winning begets winning and companies want winning imagery. Perhaps. Certainly one would not spend a lot to make an advertisement with ugly models or criminal sociopaths. But why should Shaquille O'Neal's excellence as a basketball player or announcer carry over to his choice of automobile? I might be more convinced if there wasn't a competing explanation of why this type of advertising works, but there is.

Nelson's argument is that the soft advertising does supply information, namely, the information that the company has spent, perhaps a lot, on advertising. In a nutshell, the argument is that the expenditure on advertising takes the form of an investment that can only be recouped over a long-term, profitable future for the company–a future that can only happen if the company can continue to satisfy its customers with products of appropriate quality. If the company were to cut corners on quality, its market share would dwindle and it could not recoup an appropriate rate of return on its advertising investment. The advertising investment is like posting a bond to insure future performance, and if the performance is not forthcoming, the company loses the value of the bond/investment instead of earning a profitable return on it. Simply put, the advertising assures the customer base of continued quality of service.

As a corollary of this argument, the amount spent on advertising is crucial. If not enough is spent, the seller might recoup a profitable return in the short run by selling a shoddy product once only to each buyer. Such an amount would be no assurance of quality. This also implies that the larger a company's market share and usual revenue, the larger its advertising expenditure *must* be, in the sense of having the desired effect of assuring quality, rather than *could* be, in the sense of being affordable. When advertising budgets are insufficient it brings to mind late night television, low-budget advertisements and infomercials. The actors in such commercials are likely to be employees or relatives of the proprietor rather than professional models, the production values are amateurish, and one suspects that in the case of an advertisement for jewelry, its whole cost could be recouped if the seller sold even one flawed diamond to an eager but naive young suitor.

I offer here one casual observation that supports the above interpretation. Consider how companies trumpet the news when they sign up a celebrity talent or sports superstar as a corporate spokesperson, and proudly brag about how much it cost them. Or consider how companies are happy, rather than embarrassed, about how much they pay to air a commercial during the Super Bowl. Isn't this a bit upside down? Presumably a company's shareholders are interested in the bottom line and want to procure assets of high value for lower cost rather than higher cost. Shouldn't a company's shareholders be happier about signing up Tiger Woods as a corporate spokesman and proclaiming that he likes the product so much that he will do it for free? But this rarely, if ever, happens. The company rather proclaims that Tiger Woods is endorsing the company for x years in return for y dollars per year, where x and y seem exorbitant to people of average means. The psychological argument, that excellence rubs off on the company when the company rubs elbows with celebrity talent, will work whether Tiger is paid or not, but the bonding argument only works if Tiger is paid, and works better the more he is paid.

If the reader is not persuaded by the pure theoretical argument, the question of why such advertising works will have to await further clever research for a definitive answer. There is even the possibility that both the subconscious transferal of excellence from a celebrity to the product and the economic bonding argument are both in play simultaneously. But the question does remain of whether celebrity endorsement works in the first place. To shed light on this question economists have turned to what is called an "event study."[2]

An event study uses the insights from the efficient markets view that was introduced in Chapter 11. If a stock's price quickly reflects all the public information about a company, then when new information comes out, the stock price should adjust accordingly. If the information is good, such as a decision to increase the dividend because of a higher future expected earnings profile, the stock price will jump up. If the information is bad, perhaps a government ruling to deny approval of a new drug, the

[2]See, Agrawal, J., and Kamakura, W. A. (1995). The Economic Worth of Celebrity Endorsers: An Event Study Analysis, *Journal of Marketing*, 59(3), pp. 56-62, for one of the earliest uses of an event study to look at celebrity endorsements.

stock price will fall. Of course stock prices move up and down all of the time for many reasons that might be understood after the fact even if they cannot be predicted beforehand. Perhaps most important is the tendency for stocks to move in conjunction with the whole market based upon new information that affects the economy as a whole, for example an announcement about Federal Reserve banking or interest rate policy. Event studies attempt, after the fact, to statistically correlate an individual stock's movement through time with underlying movements in the market as a whole. These movements are called the normal movements, or the normal returns, in the stock's price and can be statistically calculated with some small random error in the usual statistical sense. Then, on any one particular day, the stock's price will move idiosyncratically and the usually small statistical error will be larger on that day. This error is called the abnormal return. Event studies look at the abnormal returns on days when new information comes out about the stock.

We are finally ready to get back to golf, or golfers, or one golfer in particular, Tiger Woods. What happens in an event study when a company signs up Tiger Woods as a corporate spokesman? Tiger's excellence on the golf course led to large, even eight figure earnings, per year, from winning golf tournaments but led to about ten times as much income from endorsement contracts.[3] Besides being the highest paid professional athlete in many years of his career, Tiger's case is interesting for another less positive reason. When news of Tiger's marital infidelity broke in December, 2009, he lost some luster (and some contracts) as a celebrity endorser. We know the dates on which his endorsement deals were publicized, when news about negative personal behavior broke, and when endorsement deals were cancelled. What is left is to see what happened to the stock market prices of the relevant companies on those dates.

If it is a good decision to sign Tiger Woods as a corporate spokesman, then the abnormal returns should be positive on the day such deals were announced. If it subsequently becomes a bad decision because of Tiger's personal behavior, the stock prices should fall more than otherwise on the day(s) that bad news broke. Finally, what should happen on the day a

[3]See, http://www.forbes.com/profile/tiger-woods/.

company announces a cancelling of an endorsement contract? The abnormal returns on this final type of event could be argued to go either way. On the one hand, if the public views the company decision as incontrovertible evidence that the news about Tiger is as bad or even worse than expected, then that bad news could force the stock price down. On the other hand, if the public views the company as making a correct decision that will ultimately enhance owner value, then the stock price could rise on such a day, presumably after having already fallen due to earlier bad news. So, what is the evidence regarding these events?

Agrawal *et al.* [2012] document how firm value was significantly enhanced after endorsements by Tiger Woods. Furthermore, Knittel and Stango [2012] have examined the 10-15 trading days after the scandal broke and found that a portfolio of companies associated with Tiger Woods subsequently lost more than 2% of market value over this period. Finally, Agrawal *et al.* [2012] look at the management decision to fire or suspend advertising with Tiger Woods and found that most firms, the exception being Nike, lost additional value.

13.2 Tiger Woods and the Incentive to Supply Effort

In the contest design literature one of the central issues is how sports competition promoters can design the contest and its prize fund distributions to attract the best athletes and bring forth great effort from them.[4] Tournament-style compensation where the winners and those near the top of the ranking earn disproportionate amounts of the total payout is usually thought to provide an answer to the problem facing the promoters. Imagine, alternatively, a system where all prizes were equal, essentially equivalent to appearance fees for the "contestants." If such were the case, the best athletes might come if the fee were high enough, but no one would expect an extraordinary amount of effort to be supplied. And athletes going through the motions are not what the spectators want to see. If there is

[4]See, Szymanski, S. (2003). The Economic Design of Sporting Contests, *Journal of Economic Literature*, 41, pp. 1137–1187.

nothing at stake to winning or if the goal is not to win, then as Worf, the testosterone-laden, ultra-competitive, Clingon character from the Star Trek series, once wisely queried, "then why keep score?"

The underlying point about effort provision as a choice for professional athletes deserves further elaboration to overcome a common but very uneconomic feeling that PGA TOUR golfers (or other premier athletes for that matter) are consummate professionals who always give 100% effort. This may be a romantic vision, and may even be reinforced by sportscasters and commentators, but it is nonsense. Economic scarcity applies to the time spent training and to the level of concentration applied during a round of golf, or to any other type of effort that might be relevant, just as it applies elsewhere. Golfers could train more or concentrate more but at an added cost, which is to say at the expense of something else. If this point seems debatable, consider the question of whether it would be "easier" for a professional golfer to concentrate less and to daydream more. An economist would substitute the phrase "less costly" for "easier" but the point remains the same. If it is less costly to concentrate or train less, it is more costly to concentrate or train more.

Economically speaking, no golfers give 100% effort all the time. Neither do golfers give maximum effort. Instead, like all economic actors, golfers give "optimal" effort which is affected by the costs and benefits of effort the same as any choice that has to be made. This means that as the prize increases the benefit of effort increases and more effort is forthcoming.[5]

The efficacy of having a large prize for the winner with declining prizes as one moves down the rank order of finishers depends on the willingness of the contestants to try harder and perform better when there is more at stake. The logic of such an incentive system probably seems like common sense to most. It is also gratifying that it has been shown to be true in studies of professional golfers' performances during final rounds when it becomes clear to each contestant how much money is personally at

[5]For a further elaboration of this point as it applies to professional basketball see, Shmanske, S. (2011). Dynamic Effort, Sustainability, Myopia, and 110% Effort, *Journal of Quantitative Analysis in Sports*, 7(2).

stake.[6] Those with the most to gain or lose in the final round systematically have lower scores.

But there is still a potential problem trying to provide incentives that appears when one contestant is vastly superior to the other. The superior contestant, dubbed the "superstar," will almost always win and the opponent, seeing such a slim chance of victory, will not really try very hard. Put economically, the returns to supplying effort by the challenger are lower than the costs and less effort is supplied. Then, especially in a head-to-head competition, the superstar might also be able to coast to victory and the result is an unimpressive display of talent in the match. The possibility that the presence of a superstar will depress the effort and performance of the other competitors has motivated Jennifer Brown [2011] to consider what happens to the performance of the other competitors when Tiger Woods enters a tournament.

Brown looks at data for individual scores in PGA TOUR events and the four Majors for the years 1999-2006. A casual look at the data strongly suggests that the superstar effect is in play. First of all, Tiger Woods clearly was a superstar in these years. Tiger's average scores relative to par are always less than the average of the next 20 highest ranked golfers. In most of these years, Tiger is better by more than four strokes per tournament and in 2000 and 2006, Tiger is better by more than a whopping nine strokes per tournament. Second, by looking at individual hole scores and separating the tournaments where Woods competed from those in which he did not, Brown shows that more eagles, more birdies, fewer bogeys, and fewer double bogeys are carded when Woods is absent from the competition.

While these results are suggestive, they are not conclusive because it may simply be the case that Tiger only competes in tournaments that are staged at the most difficult golf courses. This could either be Tiger's decision about which tournaments to enter or the tournament promoter's conscious decision to try to "Tiger proof" the course. So when we look at the scores of golfers other than Tiger and find that they are higher when

[6]This has been shown statistically in two papers, [Ehrenberg and Bognanno,1990a, 1990b] and explained for non-economists in Shmanske [2004a].

Tiger competes than when he does not, we cannot tell whether they are higher because of Tiger's presence or because the golf courses are harder. This would not necessarily be a problem if there was an independent way of telling which golf courses are harder than others. If there was, then we could adjust the golfer's scores by the course difficulty and compare the adjusted scores when Tiger does or doesn't play. Unfortunately, the way that course difficulty is measured is to look at the scores that are achieved, and these are, in theory, already affected by the presence or absence of Tiger in the tournament in the first place. There would also not be a problem if all courses were of equal difficulty. A clean test comparing other golfer's scores when Tiger was in to when Tiger was out would not be complicated by saying some courses are harder than others. But, alas, this is not so. This seems like an intractable problem and it would be if it were not for Brown's cleverness, an abundance of data, and the fact that Tiger's schedule of tournaments played changed slightly from year to year.

A hint about how to solve the problem of determining whether other golfer's higher scores are due to Tiger's presence or to course difficulty is to get a way of determining the course difficulty independent of whether Tiger plays. The only assumption that Brown needs to make is that from year to year any changes in a course's underlying difficulty are negligible, especially compared to the differences in difficulty from course to course. It is only the underlying difficulty that matters because things that change from year to year like the weather (each of wind, rainfall, and temperature), can be controlled for separately. So if we assume that a particular tournament's golf course is the same from year to year, and if Tiger plays some years but not others, we have the ability to compare golfer's scores, with and without Tiger, on courses of equal difficulty. This is the basic underlying logic of Brown's research.

Thus, in a statistical regression with almost 35,000 separate scores and with controls for weather, purse size, television viewership, Major tournament status, and the golfer's ranking and usual performance at each tournament venue, Brown can isolate the effect of Tiger's presence on a golfer's scores. Significant effects, both statistically and economically, are found. For example, for players ranked in the top 20, their first round scores are an average of 0.6 strokes higher when Tiger competes. And for the tournament as a whole, for those making the cut, scores are more than 1.3 strokes higher when Tiger competes.

Brown includes many other tests, but one in particular goes to the heart of the matter concerning effort provision in the face of a superstar. These tests exploit the fact that Tiger Woods has gone through very hot streaks in his career where he seems to win multiple tournaments in a row and also through relatively colder streaks (albeit still warm by many standards) where he has changed his swing or his coach or had health or family issues and is not seen to be as dominant. According to the theory, other golfers could think they have more of a chance to win during Tiger's cooler streaks, and would be less apt to decrease their effort than when Tiger is hot. The statistics bear it out. When Tiger plays during a hot streak, other top 20 player's scores increase by almost a full stroke during the opening round, and if they make the cut, are over two strokes higher on average for the whole tournament. During Tiger's relatively cool periods he has no significant effect on other player's scores.

The importance of these findings goes beyond golf. When competitions are set up between employees for rewards, promotions, or bonuses, and the same employees always win, it could have a demoralizing effect on the other employees. A relative of mine set up a competition for "child of the week" with a $0.10 prize. When one of his four children made a commitment to win and adopted a one-upmanship strategy and won multiple weeks in a row, the other three children stopped trying. In either of these situations, a solution of having a range of rewards (second and third prizes so to speak) would seem to be a way to supply incentives to everyone in the hierarchy.

13.3 The Value of Beauty

In this section I present another example of what might be called "economic incentives where you least expect them." Seung Chan Ahn and Young Hoon Lee [2014] use LPGA tour golfers to study the relationship between beauty and earnings and have discovered a hidden or underlying linkage between the two. Perhaps unexpectedly, there is a positive link between a subjective measure of beauty and the tournament earnings of women professional golfers. Yet why should this be so in light of the fact that the tournament earnings come solely from performance on the golf course where physical appearance should have nothing to do with it? And, surprising as it may seem, this section has tangential connections to each of the previous two sections. The importance of incentives on the margin and the value of endorsement contracts each have a role to play in the connection between beauty and golfing ability.

That beauty and earnings might be related has been proposed and documented in the sociology, psychology, and economics literatures in a variety of ways.[7] The beauty standard may be subjective in and of itself, and differ by culture, but when it is identified and quantified it seems to lead to higher market returns for those with better looks. One avenue for the result is based on customer or employee discrimination through which customers and co-workers prefer to interact with good looking people. Whether or not it is fair, if TV viewers prefer handsome or pretty newscasters, then those with good looks will earn a premium in those professions.[8] The looks themselves are part of the package of productivity that a worker brings to the firm. While the part of that productivity attributable to beauty can be enhanced through surgery, grooming, and attire, the fact remains that it is the beauty itself which brings about the extra income. But one's physical appearance doesn't swing the club, so why should it matter in golf?

The bottom line answer has to do with a complementarity between beauty and golf skill in producing overall (as opposed to purely tournament) income for the golfer. Simply put, attractive golf champions will earn more money from personal appearances and endorsement contracts than average or homely golf champions, for reasons having to do with customer discrimination as mentioned above. So far so good, but now go back one step to consider the golfer's training and preparation for

[7]The pioneering work in economics is, Hamermesh, D. S. and Biddle, J. E. (1994). Beauty and the Labor Market, *American Economic Review*, 84(5), pp. 1174-94.

[8]By the way, radio is not necessarily any fairer. Looks may not be important, but those with beautiful voices will earn a premium in that medium.

tournament golf. Economically speaking, golfers have a cost-benefit decision to make concerning how much effort to put into practice and preparation. Since the after competition return is higher to the better looking, the better looking will have more incentive to practice hard and long in the first place. This effect would lead to a positive correlation between good looks and realized, as opposed to innate, golfing skill.

There is also a possible effect going in the other direction. The opportunity cost of practice might also be higher for a good looking golfer. For example, an attractive actress with innate athletic/golfing ability might find that it behooves her to pursue a career as a movie star rather than a professional golfer. This type of person would practice less leading to a negative relationship between beauty and competitive golf success. We have a classic empirical question that Ahn and Lee let the data answer.

Ahn and Lee get a subjective measure of the beauty of LPGA golfers by getting American and Asian people of varied ages and both genders to rate beauty on a one-to-five scale based on head and shoulder, and golfing follow-through pictures. Ahn and Lee are careful to choose raters unfamiliar with golf, to avoid any prejudice due to familiarity with the subjects, and to only choose pictures of golfers up to age 35. The result is a numerical beauty ranking that can be used as an independent variable in a regression setting. As an aside, this is not the first time that a subjective beauty measure has been used successfully in golf economics. However, in my research it was the beauty of the golf course that was being rated and not of the golfers.[9]

The results are interesting. Ahn and Lee look at two dependent variables, the scoring averages and the tournament earnings, for the years 1992-2010 for 116 of the 132 LPGA professional golfers who played in a Major tournament in the year 2008. For the scoring average equation, there was a nonlinear effect in which those in the lower half of the beauty ranking system are different than those in the top half. For those with above median looks, the prettier you are, the better you golf. And the effect is enhanced for those in the top 20% of the beauty ranking. For those

[9]See, Shmanske, S. (1999). The Economics of Golf Course Condition and Beauty, *Atlantic Economic Journal*, 27(3), pp. 301-313.

below the median in attractiveness, the result is in the opposite direction. The control variables in this equation include experience, which does significantly improve the scoring average, age which doesn't, and dummies for marital status and motherhood which are also not significant. One's prior expectation is presumably that there should be no correlation between beauty and golf skill, but there is a correlation. Thus the data are consistent with the posited extra incentive for attractive people to put in effort and practice. Meanwhile, in the money earnings regression, attractiveness had no extra independent effect once the scoring average and the number of tournaments played were added to the equation. This is to be expected and provides a nice robustness check on the data and the statistical methods used.

13.4 Behavioral Economics on the Golf Course

The previous two sections have identified subtly-nuanced situations in which greater potential returns are posited to lead to more effort, more time spent in preparation, and more concentration, thus leading to better performances. The statistics back up the theoretical predictions and thus lend credence to the underlying logical consistency of economic decisionmaking. Effort and concentration may fall off when competing against a superstar lowers the probability of success. Alternatively, effort and concentration will increase knowing that one's good looks will allow one to parlay tournament success into even greater earnings off the golf course. But the research reported on in this section bucks the trend of purely logical decision making by identifying a systematic difference in performance when, arguably, rationally, the stakes should be the same.

Behavioral economics is a wide-ranging field that studies the limits to and departures from purely rational decisionmaking. One such example has to do with the "framing" of certain decisions with respect to different perspectives, or "reference points" from which decisions are made. A classroom illustration might involve placing values on a common item like a coffee mug. When students do not have mugs they typically will not spend even a modest amount like a dollar to procure one. However, if free mugs are distributed to some students, those students will not part with them, even when offered a higher price of three or four dollars. From the reference point of owning the mug it seems worth more than a few dollars, but from the reference point of not owning the mug it seems worth less than a dollar. So which is worth more, the mug or the dollar? When you give students a dollar, they will not buy the mug, but when you give them the mug, they will not trade it for a dollar. Another way of saying this that is relevant to our golf example is that losing something you had is worse than not getting it in the first place, a phenomenon known as "loss aversion."

Economists Devin G. Pope and Maurice E. Schweitzer [2011] have posited that professional golfers might be influenced by loss aversion if they consider a score of par on a hole to be a relevant framing or reference point from which to assess their performance. If so, a missed par putt, which would be considered a loss from the reference point of par would hurt more than a missed birdie putt which would not be considered a loss from the reference point of par. This would be the prediction from behavioral economics, whereas the economic prediction without the behavioral economics insight would be that both missed putts would hurt exactly the same in that they would each add one stroke to the final score.

If we now fall back on regular economic logic, we would ordinarily consider the effort or concentration applied to avoiding a loss to be correlated with the size of the loss. Thus, more effort is expended when the loss to be avoided seems larger. This translates into professional golfers concentrating more on par putts than on birdie putts, and consequently making the former more often. Pope and Schweitzer, looking at over two and one half million putts hit by 421 professional golfers in 239 tournaments played from 2004 to 2009, conclude that the pros make the par putts about three percent more often than similar birdie putts. Furthermore, they calculate that the missed opportunities on the birdie putts add up to about a stroke per tournament and cost a top 20 PGA TOUR professional an average of $617,000 per year.

The Pope and Schweitzer study is made possible because for about the last decade the PGA TOUR has been using an army of volunteers and laser measuring devices to track the start and finish of every shot in every tournament. So when they claim that par putts are made more often than similar birdie putts, it is not only the distance of the putt that they measure, they also control for the direction of the putt by dividing the green into eight radial wedges each divided into long and short distances for a total of 16 different putt types. Furthermore, they do not posit a linear effect of distance on the probability of making the putt, nor a quadratic or even a cubic, but use a seventh degree polynomial to measure the nonlinear effect of distance on making the putt. This may seem like overkill, and they do not report the results of the control variables because there are simply too many, but when you have over 2.5 million observations you can afford to splurge.

I must say that I am not a fan of behavioral economics in general, but most of my quibbles with the methodology of Pope and Schweitzer seem to have been anticipated and dealt with by the authors. The main thing that comes to mind is that par putts are easier than birdie putts of the same distance because the par putt is often a follow up putt while a birdie putt is usually the first putt a golfer will hit on a hole. Related is the fact that the par putt is probably from a better angle because the previous shot was from a shorter distance than the full approach shot that set up the birdie putt. Thus, birdie putts are more likely to be randomly situated than strategically situated relative to the hole. The roll, speed, and feel of the first (birdie) putt will give information about how to hit the second (par) putt. There is also the possibility that the par putt will come after a close in chip or bunker shot on which the golfer can observe the speed and feel, whereas the birdie putt will come after a shot from 100 or more yards away which could not have been closely observed. Pope and Schweitzer attempt to control for this added information by including a control for the number of putts hit on the green by the player and his playing partners before the putt in question. Seeing other's putts is not exactly the same thing as feeling your own stroke, but this proxy, arguably, should pick up some of the effect of the added information. The fact that adding this control makes almost no difference to the main result probably indicates that the main result is not spuriously due to this omitted variable.

There is a further set of tests that lends credence to the existence of this behavioral, loss aversion, bias. Namely, the birdie putts seem to be systematically different from the par putts in the distance that they travel.

The authors are able to set up identical, or at least within one inch, pairs of putts, with one for birdie and one for par. When these putts are missed, the putts for par go two inches further than those for birdie. It seems that the putts for par are taken more aggressively while some easing of the ball up to the hole characterizes the birdie putts. This is also consistent with the do or die attitude when trying to avoid the loss of a missed par putt that is absent on birdie attempts.

All in all, Pope and Schweitzer have done an impressive job of identifying a real-world example of loss aversion bias that is economically meaningful, and that is in play even for accomplished, experienced professionals with much money at stake. This is an important result because most other examples of the phenomenon take place in artificial settings, with low stakes, untrained research subjects, and possibly biased research protocols. Unless there is something else missing from the analysis, the message for golfers is to psychologically develop the feeling that a missed birdie opportunity should hurt as much as dropping a shot to par.

13.5 The New Statistics

The study discussed in the previous section was made possible by the PGA TOUR's meticulous collection of data on every shot hit by every player. By tracking the ball from tee shot to hole out, the relative efficiency of each type of shot can be determined. Two recent papers show how the new statistics are calculated with respect to putting.[10] This section will explain the new putting statistic, called strokes gained putting, and compare the new statistic with the previous measure of putting skill in a production function regression analysis.

[10]See, Fearing, D., Acimovic, J., and Graves, S. C. (2011). How to Catch a Tiger: Understanding Putting Performance on the PGA TOUR, *Journal of Quantitative Analysis in Sports*, 7(1), Article 5, pp. 1-45, and Stockl, M., Lamb, P. F., and Lames, M. (2011). The ISOPAR Method: A New Approach to Performance Analysis in Golf, *Journal of Quantitative Analysis in Sports*, 7(1), Article 10, pp. 1-15.

To understand the new putting statistic, consider a tap in putt of the length that was made 100% of the time on a particular green. Each golfer would take one stroke to hole out and the average of all golfers is to take one stroke to hole out. Since the golfer takes one stroke to do what every other golfer takes one stroke to do the golfer neither gains nor loses a stroke on that putt. Now consider a putt that was made only 50% of the time, and when missed, left a simple tap in putt. Half of the golfers took two shots to finish the hole but the other half only took one stroke. So a golfer who made the putt took one stroke which took, on average, 1.5 strokes, so the golfer gained 0.5 strokes on that putt. Meanwhile, a golfer who took two shots to hole out lost 0.5 strokes putting on that green. These gains and losses are summed for all eighteen greens to get the "strokes gained putting" statistic. This method controls for the distance and directional difficulty of putting much more closely than simply counting the average number of putts taken on greens hit in regulation, which was the measure previously used to measure the golf professional's putting skill.

To compare the precision of the different putting statistics I estimate the following production and earnings function with data from the PGA TOUR 2012 season:

$$Y = b_0 + b_1 DRIVDIST + b_2 DRIVACC + b_3 APPROACH + b_4 PUTTING + b_5 SANDSAVE + e . \quad (13.1)$$

Two dependent variables will be examined, SCOREAVE, which is the scoring average of the golfer for the season, and LNYPERTO, which is the natural logarithm of earnings per tournament for the golfer.[11] These are the typical variables addressed in many studies of the relationship between skills and golfer earnings or between skills and performance, and will allow comparison of these results with previous results especially those highlighted or employed in Chapters 5, 6, 7, and 9 of this book. When SCOREAVE is used the equation is a production function in which the output, SCOREAVE, is a function of the inputs, namely the five skills in

[11]When the level of money earnings is used instead of the logarithm the results are the same in essence. For statistical reasons using the logarithm yields better estimates.

Eq. (13.1). When LNYPERTO is used the equation is a reduced-form earnings function in which skills produce scores and scores produce winnings, but the scores are suppressed and the winnings depend directly upon the skills. Both methods have been used in the past and have been explained fully in Chapter 6.

With respect to the independent variables, DRIVDIST and DRIVACC are the distance in yards of the golfer's drives and the percentage of times the drives end up in the fairway. These two measures of the skill of driving have been used in just about every study of the golf production function. SANDSAVE is also the usual measure of the percentage of times that a player can finish the hole in two or fewer shots from a greenside bunker. Most of our attention will be paid to the approach shots and the putting, each of which will be measured in two ways. GIR is the usual way of measuring accuracy with approach shots and is the percentage of times that the golfer reaches the green in regulation, meaning being on the putting surface in two shots less than par. Approach shot accuracy will also be measured by the presumably superior DISTFRHO, which stands for the distance from the hole that the approach shot finishes and is made possible by the laser accuracy of the PGA TOUR's new statistical measuring efforts. With respect to putting, PUTTPER is the average number of putts taken on greens hit in regulation, and STRSPUTT is the new measure of strokes saved putting as explained above. The summary statistics for all the variables used are in Table 13.1.

Before actually running the regressions I was curious to find out how the new putting and approach shot accuracy measures compared to the old ones. For putting, the simple correlation coefficient was -0.67. We expect the negative sign of course, because PUTTPER is measured in strokes per green and STRSPUTT is measured in strokes *saved* per round. This correlation is fairly close, in fact, closer than any other pairwise comparison of variables. However, it is not so close as to count out any pertinent difference between the two. That is to say, the STRSPUTT theoretically should be the better measure, but even so, if the two were very closely correlated it might not even make a difference. The simple correlation between GIR and DISTFRHO is -0.50. Again, we expect a negative correlation because we want to hit more greens so increases in

Table 13.1 Summary statistics.

Variable	Mean	Std. Dev.	Minimum	Maximum	N
SCOREAVE	70.89	0.71	68.87	73.00	191
YPERTO	60.72	65.08	2.07	503.00	190
PUTTPER	1.78	0.0233	1.718	1.841	191
STRSPUTT	0.0167	0.338	-1.177	0.86	191
GIR	64.88	2.70	57.74	70.34	191
DISTFRHO	35.90	1.48	32.10	39.90	191
DRIVDIST	290.05	8.38	268.9	315.5	191
DRIVACC	61.03	4.74	47.27	73.00	191
SANDSAVE	48.58	6.21	29.73	65.44	191

Notes: SCOREAVE, PUTTPER, and STRSPUTT are in strokes. YPERTO is in 1,000's of 2012 dollars. GIR, DRIVACC, and SANDSAVE are percentages times 100. DISTFRHO is in feet. DRIVDIST is in yards. Source: www.pgatour.com and author's calculations.

GIR are good, but increases in DISTFRHO, the distance to the hole, are bad. Nothing else jumps out from the summary statistics which are in line with previous years. The only skill that really has changed over the years is driving distance as discussed in Chapter 8. At right around 290 yards, the average driving distance has been level since about 2005.

At this point I fully expected to present the results showing the superiority of the new putting measure and call it a day. Alas, unexpectedly, the old putting measure works as well, and sometimes better, than the new one in my regressions. I consider this to be a puzzle in need of an explanation so I will offer readers the following challenge. For the best explanation of what is going on with my comparison of PUTTPER and STRSPUTT, received by December 31, 2015, I will send a signed copy of this book. Big woop!

Let us turn to the results which are listed in Tables 13.2 and 13.3. Table 13.2 lists the results of the SCOREAVE equation. The first column of results, which I will call the "old" results, uses the older PUTTPER and GIR measurements of putting and approach shot accuracy. The next two

Table 13.2 Regressions of scoring average on skills, (t-statistics).

Dep. Variable	SCOREAVE	SCOREAVE	SCOREAVE	SCOREAVE	SCOREAVE
Constant	69.13 (22.5)	95.41 (61.2)	69.51 (20.6)	88.05 (41.8)	74.56 (20.58)
PUTTPER	12.83 (9.54)		10.47 (7.45)		8.20 (4.74)
STRSPUTT		-0.762 (-7.57)		-0.747 (-7.50)	-0.442 (-3.69)
GIR	-0.090 (-6.36)	-0.084 (-5.61)			-0.079 (-5.44)
DISTFRHO			0.091 (3.78)	0.141 (5.80)	0.073 (3.00)
DRIVDIST	-0.035 (-7.30)	-0.045 (-8.96)	-0.050 (-11.56)	-0.055 (-12.94)	-0.039 (-8.17)
DRIVACC	-0.054 (-6.09)	-0.068 (-7.18)	-0.073 (-8.38)	-0.073 (-8.42)	-0.050 (-5.59)
SANDSAVE	-0.034 (-6.88)	-0.037 (-6.99)	-0.034 (-6.44)	-0.034 (-6.32)	-0.031 (-6.31)
R-squared	0.715	0.675	0.677	0.679	0.740
Adj. R-squared	0.707	0.666	0.669	0.670	0.730
N	191	191	191	191	191

Note: All variables significant at 0.01 level.

columns replace one or the other of these with their newer versions, STRSPUTT and DISTFRHO. The fourth column contains the "new" results using both new measures simultaneously. Finally, the fifth column includes both measures of each of putting and approach shot accuracy. All the equations also control for driving distance and accuracy, and greenside bunker skills.

Every variable in every equation is statistically significant with a p-value of 0.01 or below with the right sign and reasonable magnitude. For example, to save a stroke by driving the ball farther you need somewhere between 18 (the coefficient estimate of -0.055) and 29 (the coefficient estimate of -0.035) yards of extra distance. To save a stroke by hitting more greens in regulation, consider the estimate of -0.090 in the first column which means that for each percentage point increase in GIR, the average score falls by 0.09 strokes. Since 11 times 0.09 is approximately one, an improvement of 11 percentage points in GIR will save a stroke per round. On average, the professionals currently hit about 64.88% of the greens for an average of 11.7 greens per round. An increase of 11 percentage points to 75.88% means hitting an average of 13.7 greens per round. That this would save a stroke per round or four strokes per tournament is very believable as a point estimate. By the way, this improvement would be about four standard deviations from the mean and none of the golfers in 2012 reached this level of performance. Consider just one more back of the envelope type calculation. How much more accurate in closeness to the hole is required to save a stroke per round. Dividing one (stroke) by the point estimate of 0.141 strokes per feet indicates that an improvement of about seven feet is required to save a stroke per round. Similarly to the calculation using the GIR coefficient, an improvement of seven feet closer to the hole is over a four standard deviation improvement that none of the golfers achieved in the 2012 data.

There is little to quarrel with as to these results except that the newer, theoretically improved statistics do not work better than the old ones. If anything, they are a little worse, but the differences are negligible. Theoretically, the PUTTPER measure only considers whether or not the approach shot is on the green and does not control for the distance of the putts. Therefore, it measures the skill of putting with more error than the STRSPUTT variable. If this were the only difference, its coefficient is biased toward zero and it should not add as much explanatory power as STRSPUTT. It must be the case, therefore, that the PUTTPER variable is also capturing something else that is important to scoring. My tentative

Table 13.3 Regressions of earnings on skills, (t-statistics).

Dep. Variable	LNYPERTO	LNYPERTO	LNYPERTO	LNYPERTO	LNYPERTO
Constant	16.02 (2.98)	-20.01 (-7.48)	15.75 (2.80)	-11.44 (-3.19)	8.54 (1.31)
PUTTPER	-17.66 (-7.46)		-15.49 (-6.56)		-11.87 (-3.77)
STRSPUTT		1.030 (5.98)		1.046 (6.23)	0.557 (2.57)
GIR	0.078 (3.13)	0.068 (2.64)			0.066 (2.51)
DISTFRHO			-0.082 (-2.04)	-0.156 (-3.79)	-0.078 (-1.76)
DRIVDIST	0.051 (6.05)	0.065 (7.50)	0.063 (8.80)	0.071 (9.81)	0.055 (6.52)
DRIVACC	0.077 (4.92)	0.094 (5.85)	0.092 (6.21)	0.090 (6.05)	0.073 (4.51)
SANDSAVE	0.028 (3.23)	0.033 (3.59)	0.028 (3.20)	0.029 (3.25)	0.024 (2.74)
R-squared	0.540	0.499	0.526	0.517	0.559
Adj. R-squared	0.528	0.485	0.513	0.504	0.542
N	190	190	190	190	190

Note: 30 of 32 estimated coefficients are significant at 0.01 level.

guess is that PUTTPER includes some information about the quality of the approach shot leading to the birdie pùtt that is not already being captured by either (or both) of GIR and DISTFRHO. This suspicion is consistent with the fact that the significance of PUTTPER goes down but does not completely disappear even when all of STRSPUTT, GIR, and DISTFRHO are included. However, what this possible source of extra information is will remain a mystery until some clever reader explains it to me.

We may briefly turn our attention to Table 13.3 in which LNYPERTO, the natural logarithm of the money earned per tournament, is the dependent variable. Table 13.3 is organized in the same fashion as Table 12.2. The statistical character of the results are good. Except for one constant term and DISTFRHO in a couple of equations all the coefficients are tightly estimated and significant at the 99% level, and DISTFRHO is significant at the 90% and 95% levels in the offending cases. The, percentage of the variation explained in the model is right around 50% as indicated by the R-squared figures. This compares favorably with identical specifications estimated on other years of data. For example, in Chapters 6 and 7, in Tables 6.2, 7.4, and 7.5, for the years 2006, 1998, and 2008, the adjusted R-squared figures, respectively, are 0.362, 0.516, and 0.203.

Again we see the tendency for each combination of explanatory variables to tell the same story, with the older formulation of the putting and approach shot variables leading to the best results. With respect to the coefficient estimates themselves, consider the entry in the first column for GIR. The coefficient estimates are interpreted as percentage changes when the dependent variable is transformed to its logarithm. Therefore, the 0.078 figure for GIR means that increasing the number of greens hit in regulation by one percentage point will lead to an increase in earnings of 7.8%. Meanwhile, the estimated coefficient for DRIVDIST is 0.051 meaning that one more yard of distance corresponds to a 5.1% increase on average in earnings per tournament.

The putting variable is still, by far, the most important. In the first column the estimate for PUTTPER is -17.66. This huge effect comes from an increase of one putt for every green reached in regulation, which is about 47 or 48 strokes for a four-day tournament. Dividing the effect by 48 to get the effect of one stroke per tournament yields an increase in earnings per tournament of 36.8%. This would seem very reasonable. At the top of the leaderboard, one stroke saved could almost double the prize. For the average earnings in our sample from Table 13.1, a 36.8% increase corresponds to an increase of over $22,000 per tournament.

Let us compare this to what the new strokes saved putting measure says. In the second column, the coefficient of STRSPUTT is 1.03. One stroke saved *per eighteen holes* increases earnings by 103%. To get the effect for one stroke saved per tournament we must divide by four. At the average earnings of $60,720 a 25.75% increase corresponds to over $15,600 per tournament, and in the same ballpark, or should I say fairway, as the amount derived from the estimates using the older PUTTPER variable.

I'll abruptly leave you here, hopefully wanting more. The data is available and the gauntlet has been thrown in my challenge to explain why STRSPUTT doesn't work better. I will certainly be thinking about in my retirement years, especially as I recount whether my putting has catapulted me or kept me from hitting my handicap in my most recent round.

References

1995 USGA Pace Rating System, (United States Golf Association, Far Hills, New Jersey), 1995.

2003 PGA TOUR Media Guide, PGA TOUR, Ponte Vedra Beach, Florida.

2008 U.S. Open Economic Impact Analysis, San Diego State University, Center for Hospitality and Tourism Research.

Agrawal, J., and Kamakura, W. A. (1995). The Economic Worth of Celebrity Endorsers: An Event Study Analysis, *Journal of Marketing*, 59(3), pp. 56-62.

Agrawal, J., Grimm, P. and Fung, S. (2013). Benefits and Costs of Hiring and Firing Tiger Woods, California State University, East Bay, Working Paper, December.

Ahn, S. C. and Lee, Y. H. (2014). Beauty and Productivity: The Case of the Ladies Professional Golf Association, *Contemporary Economic Policy*, 32(1), pp. 155-168.

Alchian, A. (1974). Information, Martingales and Prices, *Swedish Journal of Economics*, 76(1), pp. 3-11.

Alexander, D. L. and Kern, W. (2005). Drive for Show and Putt for Dough?: An Analysis of the Earnings of PGA Tour Golfers, *Journal of Sports Economics*, 6(1), pp. 46-60.

Andriot, J. L. ed. (1983). *Population Abstract of the United States*, (Andriot Associates, McLean, VA).

von Allmen, P. (2012). *The Oxford Handbook of Sports Economics, Volume 2: Economics Through Sports*, eds. Shmanske, S. and Kahane, L. H. Chapter 18 "Multiplier Effects and Local Economic Impact," (The Oxford University Press, New York) pp. 321-334.

Asch P., Malkiel, B. G., and Quandt. R. E. (1982). Racetrack Betting and Informed Behavior, *Journal of Financial Economics*, 10, pp. 187-194.

Baade, R. and Matheson V. (2000). An Assessment of the Economic Impact of the American Football Championship, the Super Bowl, on Host Communities, *Reflets et Perspectives*, 30, pp. 35-46.

Barzel, Y. (1974). A Theory of Rationing by Waiting, *Journal of Law and Economics*, 17, pp. 73-95.

Berry, S. M. (1999). Drive for Show and Putt for Dough, *Chance*, 12(4), pp. 50-55.

Better, C. (2012). How to get Tee Times on Bethpage Black Golf Course, Site of the 2002 and 2009 US Opens, accessed at http://www.golfvacationinsider.com/cr/bethpage-black-golf-course, on May 21, 2012.

Blofsky, E. T. Jr. ed. (2002). *NCGA Golf 2002 Bluebook Edition*, Vol. 22, No. 1.

Branch, J. (2009). Parking All Night at Bethpage, Hoping to Drive, *New York Times*, June 6.

Brown, J. (2011). Quitters Never Win: the (Adverse) Incentive Effects of Competing with Superstars, *Journal of Political Economy*, 119(5), pp. 982-1013.

Callan, S. J. and Thomas, J. M. (2007). Modeling the Determinants of a Professional Golfer's Tournament Earnings: A Multiequation Approach, *Journal of Sports Economics*, 8(4), pp. 394-411.

Camerer, C. F. (1989). Does the basketball market believe in the 'hot hand'?, *American Economic Review*, 79(5), pp. 1257-61.

Coates, D. and Humphries, B. R. (2008). Do Economists Reach a Conclusion on Subsidies for Sports Franchises, Stadiums, and Mega-Events?, *Econ Journal Watch*, 5(3), pp. 294-315.

Connolly, R. A. and Rendleman Jr., R. J. (2008). Skill, Luck, and Streaky Play on the PGA Tour, *Journal of the American Statistical Association*, 103(481), pp. 74-88.

Davidson, J. D. and Templin, T. J. (1986). Determinants of Success Among Professional Golfers, *Research Quarterly for Exercise and Sport*, 57(1), pp. 60-67.

Demsetz, H. (1968). The Cost of Transacting, *Quarterly Journal of Economics*, 82, pp. 33-53.

Ehrenberg, R. G. and Bognanno, M. L. (1990a). Do tournaments have incentive effects?, *Journal of Political Economy*, 98(6), 307-24.

Ehrenberg, R. G. and Bognanno, M. L. (1990b). The incentive effects of tournaments revisited: Evidence from the European PGA Tour, *Industrial and Labor Relations Review*, 43, pp. 74S-88S.

Fearing, D., Acimovic, J., and Graves, S. C. (2011). How to Catch a Tiger: Understanding Putting Performance on the PGA TOUR, *Journal of Quantitative Analysis in Sports*, 7(1), Article 5, pp. 1-45.

Fitzgerald, T. (June 20, 2006). Sports TV: Without Tiger, U.S. Open ratings tank. *medialifemagazine.com*, accessed at: http://www.medialifemagazine.com/cgi-bin/artman/exec/view.cgi?archive=238&num=5473.

Fort, R. (2003). Thinking (some more) about competitive balance, *Journal of Sports Economics*, 4(4), pp. 280-3.

Fried, H. O. Lambrinos, J., and Tyner, J. (2004). Evaluating the Performance of Professional Golfers on the PGA, LPGA, and SPGA Tours, *European Journal of Operational Reseach*, 154(2), pp. 548-561.

Gilley, O. W. and Chopin, M. C. (2000). Professional Golf: Labor or Leisure, *Managerial Finance*, 26(7), pp. 33-45.

Golf Consumer Profile 1989 Edition, National Golf Foundation, August 1989.

Granger, C. W. J. (1969). Investigating Causal Relations by Econometric Models and Cross-Spectral Methods, *Econometrica*, 37(3), pp. 424-438.

Graves, R. M. and Cornish, G. S. (1998). *Golf Course Design*, (John Wiley & Sons, Inc., New York).

Hagn, F. and Maennig, W. (2008). Employment Effects of the Football World Cup 1974 in Germany, *Labour Economics*, 15(5), pp. 1062-1075.

Hamermesh, D. S. and Biddle, J. E. (1994). Beauty and the Labor Market, *American Economic Review*, 84(5), pp. 1174-94.

Hood, M. (2006). The Purse is not Enough: Modeling Professional Golfers Entry Decision, *Journal of Sports Economics*, 7(3), pp. 289-308.

Hotchkiss, J. L., Moore, R. E. and Zobay, S. M. (2003). Impact of the 1996 Summer Olympic Games on Employment and Wages in Georgia, *Southern Economic Journal*, 69(3), pp. 691-704.

Hotelling, H. (1929). Stability in Competition, *Economic Journal*, 39, pp. 41-57.

Humphreys, B. R. (2002). Alternative measures of competitive balance in sports leagues, *Journal of Sports Economics*, 3(2), pp. 133-48.

Hurley, W. and McDonough, L. (1995). A Note on the Hayek Hypothesis and the Favorite-Longshot Bias in Parimutuel Betting, *American Economic Review*, 85(4), pp. 949-955.

Kahane, L. H. (2003). Comments on 'Thinking about competitive balance,' *Journal of Sports Economics*, 4(4), pp. 288-91.

Kahane, L. H. (2010). Returns to Skills in Professional Golf: A Quantile Regression Approach, *International Journal of Sport Finance*, 5(3), pp. 167-180.

Kahane, L. H. and Shmanske, S. eds. (2012). *The Oxford Handbook of Sports Economics, Volume 1: The Economics of Sports*, (Oxford University Press, Inc., New York).

Kamer, P. M. (2009). The 2009 U.S. Golf Open at Bethpage Black: Its Impact on the Long Island Economy, Research Report from the Long Island Association.

Knittel, C. R. and Stango, V. (2012). Celebrity Endorsements, Firm Value, and Reputation Risk: Evidence from the Tiger Woods Scandal, *Management Science, Articles in Advance,* pp. 1-17.

Kuper, S. and Szymanski, S. (2009). *Soccernomics: Why England Loses, Why Spain, Germany, and Brazil Win, and Why the US, Japan, Australia, Turkey-and Even Iraq-Are Destined to Become the Kings of the World's Most Popular Sport*, (Nation Books, New York).

Lazear, E. P. and Rosen, S. (1981). Rank-order tournaments as optimum labor contracts, *Journal of Political Economy*, 89(5), 841-64.

Levitt, S. D. and Dubner, S. J. (2005). *Freakonomics: A Rogue Economist Explores the Hidden Side of Everything*. (William Morrow an imprint of HarperCollins, New York).

Levitt, S. D. and Dubner, S. J. (2009). *SuperFreakonomics: Global Cooling, Patriotic Prostitutes, and Why Suicide Bombers Should Buy Life Insurance*. (HarperCollins Publishers, New York).

Livingstone, S. (2009). Tee Time at Bethpage Worth the Wait . . . Overnight in Parking Lot, *USA Today*, June 18.

Marketing Plan Manual, National Golf Course Owners Association, 1987.

Matheson, V. A. (2012). *The Oxford Handbook of Sports Economics, Volume 1: The Economics of Sports*, eds. Kahane, L. H. and Shmanske, S., Chapter 24 "Economics of the Super Bowl," (The Oxford University Press, New York) pp. 470-84.

Moy, R. L. and Liaw, T. (1998). Determinants of professional golf tournament earnings, *The American Economist*, 42(1), pp. 65-70.

National Football League (1999). "Super Bowl XXXIII generates $396 million for South Florida," NFL Report, 58.

NCGA Bluebook 2002 edition.

Nelson, P. (1974). Advertising as Information, *Journal of Political Economy*, 82(4), pp. 729-754.

Nero, P. (2001). Relative salary efficiency of PGA Tour golfers, *The American Economist*, 45(1), pp. 51-6.

Newport, J. P. (2013). The Real Causes of Slow Play, *Wall Street Journal*, July 13-14.

Nix, C. L. and Koslow, R. (1991). Physical Skill Factors Contributing to Success on the Professional Golf Tour, *Perceptual and Motor Skills*, 72, pp. 1272-1274.

Noll, R G. and Zimbalist, A. eds. (1997). *Sports, Jobs, and Taxes*. (Brookings Institution Press, Washington, DC).

Oaxaca, R. (1973). Male-Female Wage Differentials in Urban Labor Markets, *International Economic Review*, 14, pp. 693-709.

Oaxaca, R., and Ransom, M. R. (1999). Identification in Detailed Wage Decompositions, *The Review of Economics and Statistics*, 81, pp. 154-157.

Pepall, L., Richards, D. J., and Norman, G. (2002). *Industrial Organization: Contemporary Theory and Practice*, (Southwestern Publishing, Oklahoma City) chapters 4 and 8.

Pfitzner, C. B. and Rishel, T. D. (2005). Performance and Compensation on the LPGA Tour: A Statistical Analysis, *International Journal of Performance Analysis in Sport*, 5(3) pp. 29-39.

Pope, D. G. and Schweitzer, M. E. (2011). Is Tiger Woods Loss Averse? Persistent Bias in the Face of Experience, Competition, and High Stakes, *American Economic Review*, 101(1), pp. 129-57.

Rhoads, T. A. (2007a). Labor Supply on the PGA TOUR: The Effect of Higher Expected Earnings and Stricter Exemption Status on Annual Entry Decisions, *Journal of Sports Economics*, 8(1), pp. 83-98.

Rhoads, T. A. (2007b). On Nonlinear Prizes in Professional Golf. conference paper presented at WEAI meetings in Seattle, July.

Rishe, P. J. (2001). Differing rates of return to performance: A comparison of the PGA and Senior golf tours, *Journal of Sports Economics*, 2(3), pp. 285-96.

Rosen, S. (1981). The economics of superstars, *American Economic Review*, 71(5), pp. 845-58.

Sanderson, A. R. and Siegfried, J. J. (2003). Thinking about competitive balance, *Journal of Sports Economics*, 4(4), pp. 255-79.

Sauer, R. D. (1998). The Economics of Wagering Markets, *Journal of Economic Literature*. 36(4), pp. 2021-2064.

Schmidt, M. B. and Berri, D. J. (2001). Competitive balance and attendance: The case of Major League Baseball, *Journal of Sports Economics*, 2(2), pp. 145-67.

Scully, G. W. (2002). The Distribution of Performance and Earnings in a Prize Economy, *Journal of Sports Economics*, 3(3), pp. 235-45.

Shaffer, T. L., Connaughton, D. P., Siders, R. A., and Mahoney, J. F. (2000). An Analysis of the Most Significant Variables for Predicting Scoring Average and Money Won per Event in Professional Golf, *Research Quarterly for Exercise and Sport*, 71(Supp.), pp. 119A-120A.

Shmanske, S. (1991). Tied Bets, Half Points, and Price Discrimination, *Kentucky Journal of Economics and Business*, 11, pp.43-54.

Shmanske, S. (1992). Human capital formation in professional sports: Evidence from the PGA Tour, *Atlantic Economic Journal*, 20(3), pp. 66-80.

Shmanske, S. (1996). Contestability, Queues, and Governmental Entry Deterrence, *Public Choice*, 86, pp. 1-15.

Shmanske, S. (1998a). Price Discrimination at the Links, *Contemporary Economic Policy*, 16(3), pp. 368-378.

Shmanske, S. (1998b). Subjective Measurement and 'Bad-Mood' Bias, *Briefing Notes in Economics*, 35, pp. 1-4.

Shmanske, S. (1999). The Economics of Golf Course Condition and Beauty, *Atlantic Economic Journal*, 27(3), pp. 301-313.

Shmanske, S. (2000). Gender, Skill, and Earnings in Professional Golf, *Journal of Sports Economics*, 1(4), pp. 385-400.

Shmanske, S. (2004a). *Golfonomics*. (World Scientific Publishing Co., Inc. River Edge, NJ).

Shmanske, S. (2004b). Market Preemption and Entry Deterrence: Evidence from the Golf Course Industry, *International Journal of the Economics of Business*, 11, pp. 55-68.

Shmanske, S. (2005). Odds-Setting Efficiency in Gambling Markets: Evidence from the PGA TOUR, *Journal of Economics and Finance*, 29(3), pp. 391-402.

Shmanske, S. (2007). Consistency or Heroics: Skewness, Performance and Earnings on the PGA TOUR, *The Atlantic Economic Journal*, 35(4), pp. 463-471.

Shmanske, S. (2008). Skills, Performance, and Earnings in the Tournament Compensation Model: Evidence from PGA TOUR Microdata, *Journal of Sports Economics*, 9(6), pp. 644-62.

Shmanske, S. (2009). Golf Match: The Choice by PGA Tour Golfers of Which Tournaments to Enter, *International Journal of Sports Finance*, 4(2), pp. 114-135.

Shmanske, S. (2011). Dynamic Effort, Sustainability, Myopia, and 110% Effort, *Journal of Quantitative Analysis in Sports*, 7(2).

Shmanske, S. (2012a). *Handbook on the Economics of Mega Sporting Events*, eds. Maennig, W. and Zimbalist, A., Chapter 25 "The Economic Impact of the Golf Majors," (Edward Elgar Publishing Ltd, Northampton, Massachusetts) pp. 449-460.

Shmanske, S. (2012b). *The Oxford Handbook of Sports Economics, Volume 2: Economics through Sports*, eds. Shmanske, S. and Kahane, L. H., Chapter 3 "Gender and Discrimination in Professional Golf," (Oxford University Press, Inc., New York) pp. 39-54.

Shmanske, S. (2013). *Handbook on the Economics of Women's Sports*, eds. Leeds, E. M. and Leeds, M. A., Chapter 4 "Gender and Skill Convergence in Professional Golf," (Edward Elgar Publishing Ltd, Northampton, Massachusetts) pp. 73-91.

Shmanske, S. and Kahane, L. H. eds. (2012). *The Oxford Handbook of Sports Economics, Volume 2: Economics through Sports*, (Oxford University Press, Inc., New York).

Sommers, P. M. (1994). A Bread and Putter Model, *Atlantic Economic Journal*, 22, p. 77.

Stachura, M. (2002). About-Face: The USGA's Final Edict on COR Should End the Confusion Over Which Drivers Conform and Which Do Not, *Golf Digest*, 53(October).

Stockl, M., Lamb, P. F., and Lames, M. (2011). The ISOPAR Method: A New Approach to Performance Analysis in Golf, *Journal of Quantitative Analysis in Sports*, 7(1), Article 10, pp. 1-15.

Stone, D. F. (2012). Measurement Error and the Hot Hand, WEAI 2012 Conference paper.

Szymanski, S. (2001). Income inequality, competitive balance, and the attractiveness of team sports: Some evidence and a natural experiment from English soccer, *The Economic Journal*, 111(469), pp. F69-F84.

Szymanski, S. (2003). The Economic Design of Sporting Contests, *Journal of Economic Literature*, 41, pp. 1137–1187.

Thaler, R. H. and Ziemba, W. T. (1988). Parimutuel Betting Markets: Racetracks and Lotteries, *Journal of Economic Perspectives*, 2(2), pp. 161-174.

Wiseman, F., Chatterjee, S., Wiseman, D., and Chatterjee, N. (1994). *Science and Golf II. Proceedings of the World Scientific Congress of Golf*, eds. Cochran, A. J. and Farrally, M. R. "An Analysis of 1992 Performance Statistics for Players on the U.S. PGA, Senior PGA and LPGA Tours," (E&FN SPON, London), pp. 199-204.

White, H. (1980). A Heteroskedastic-Consistent Covariance Matrix Estimator and a Direct Test for Heteroskedasticity, *Econometrica*, 48(4), pp. 817-38.

Zimbalist, A. S. (2002). Competitive balance in sports leagues: An introduction, *Journal of Sports Economics*, 3(2), pp. 111-21.

Zuber, R., Gandar, J. M. and Bowers, B. (1985). Beating the Spread: Testing the Efficiency of the Gambling Market for NFL Games, *Journal of Political Economy*, 93(4) pp. 800-806.

http://www.bea.gov/bea/regional/rims/.

http://www.census.gov/econ/cbp/index.html.

http://www.forbes.com/profile/tiger-woods/.

http://www.golfvacationinsider.com/cr/bethpage-black-golf-course, on May 21, 2012.

http://www.lpga.com.

http://www.medialifemagazine.com/cgi-bin/artman/exec/view.cgi?archive=238&num=5473.

http://www.ngf.org/faq/growthofgolf.html.

http://www.pay-equity .org/info-time.html, accessed on 8/1/2009.

http://www.pgatour.com.

Index

Printed in the United States
By Bookmasters